SPPU New Syllabus

Z0122

CELL BIOLOGY

[2 Credits]

For

F.Y.B.Sc. Zoology, Paper – II : Semester II

As Per New Revised Syllabus, CBCS Pattern

From June 2019

PRIN. DR. KISHORE R. PAWAR
M.Sc., Ph.D.
Former Principal,
Karmveer Shantarambapu Kondaji Wavare,
Arts, Sci. & Com. College, CIDCO,
NASHIK – 422008.

DR. ASHOK E. DESAI
M.Sc., Ph.D.
Former Associate Professor &
Head, P.G. Deptt. of Zoology,
K.T.H.M. College,
NASHIK – 422002.

NIRALI PRAKASHAN
ADVANCEMENT OF KNOWLEDGE

N5035

F.Y.B.Sc. CELL BIOLOGY **ISBN 978-93-89533-62-0**

First Edition : November 2019

Published By:

NIRALI PRAKASHAN

Abhyudaya Pragati, 1312, Shivaji Nagar
Off J.M. Road, PUNE – 411005
Tel - (020) 25512336/37/39, Fax - (020) 25511379
Email : niralipune@pragationline.com

➢ DISTRIBUTION CENTRES

PUNE

Nirali Prakashan : (For orders within Pune) 119, Budhwar Peth, Jogeshwari Mandir Lane, Pune 411002, Maharashtra, Tel : (020) 2445 2044, Mobile : 9657703145
Email : niralilocal@pragationline.com

Nirali Prakashan : (For orders outside Pune) S. No. 28/27, Dhayari, Near Asian College Pune 411041
Tel : (020) 24690204; Mobile : 9657703143
Email : bookorder@pragationline.com

MUMBAI

Nirali Prakashan : 385, S.V.P. Road, Rasdhara Co-op. Hsg. Society Ltd., Girgaum, Mumbai 400004, Maharashtra;
Mobile : 9320129587 Tel : (022) 2385 6339 / 2386 9976, Fax : (022) 2386 9976
Email : niralimumbai@pragationline.com

➢ DISTRIBUTION BRANCHES

JALGAON

Nirali Prakashan : 34, V. V. Golani Market, Navi Peth, Jalgaon 425001, Maharashtra, Tel : (0257) 222 0395, Mob : 94234 91860;
Email : niralijalgaon@pragationline.com

KOLHAPUR

Nirali Prakashan : New Mahadvar Road, Kedar Plaza, 1st Floor Opp. IDBI Bank, Kolhapur 416 012, Maharashtra. Mob : 9850046155;
Email : niralikolhapur@pragationline.com

NAGPUR

Nirali Prakashan : Above Maratha Mandir, Shop No. 3, First Floor, Rani Jhanshi Square, Sitabuldi, Nagpur 440012, Maharashtra
Tel : (0712) 254 7129;
Email : niralinagpur@pragationline.com

DELHI

Nirali Prakashan : 4593/15, Basement, Agarwal Lane, Ansari Road, Daryaganj Near Times of India Building, New Delhi 110002
Mob : 08505972553, Email : niralidelhi@pragationline.com

BENGALURU

Nirali Prakashan : Maitri Ground Floor, Jaya Apartments, No. 99, 6th Cross, 6th Main, Malleswaram, Bengaluru 560003, Karnataka;
Mob : 9449043034
Email: niralibangalore@pragationline.com

Other Branches : Hyderabad, Chennai

niralipune@pragationline.com | www.pragationline.com

Also find us on f www.facebook.com/niralibooks

Preface ...

We have great pleasure in presenting this book **'A Textbook of Cell Biology'** for the F. Y. B. Sc. Zoology students of Savitribai Phule Pune University, Pune.

The present book is written as per the revised syllabus with effect from June 2019. The main objective of writing this book is to present the subject matter in a compiled, concise and easily comprehensive form keeping in mind the basic needs of the students.

This text book has been written in a simple and lucid language. Several illustrative figures and diagrams have been included in the text which will help the students to grasp the concepts quickly and easily. We strongly believe that this book will be of great help to students as well as teachers.

We express our sincere thanks to Mr. Dineshbhai Furia, Mr. Jignesh Furia and the staff of Nirali Prakashan for bringing out this book in time.

We would appreciate suggestions and feedback from colleagues and students for further improvement of the book.

November, 2019

Prin. Dr. Kishore R. Pawar

Dr. Ashok E. Desai

Syllabus ...

1. **Introduction:** **(04)**
   1.1 Introduction to cell biology.
   1.2 Cell as basic unit of life.
   1.3 Importance of Cell Biology and its applications in industry.

   Overview of Cells

   1.3 Introduction to Prokaryotic and Eukaryotic cells.
   1.4 Structure and function of Prokaryotic (*E. coli*).
   1.5 Structure and function of Eukaryotic cells (Animal and Plant Cell).
2. **Techniques in Cell Biology** **(04)**
   2.1 Introduction
   2.2 Microscopy: Basic Principle, Simple, Compound and applications of Electron Microscope.
   2.3 Stains and dyes:
   Types of Stain: Acidic, basic and neutral.
   Dye (Preparation and chemistry of dyes not expected)
   2.4 Micrometry.
3. **Plasma Membrane** **(06)**
   3.1 Introduction
   3.2 Structure of plasma membrane: Fluid mosaic model.
   3.3 Transport across membranes: Active and Passive transport, Facilitated transport, exocytosis, endocytosis, phagocytosis – vesicles and their importance in transport.
   3.4 Other functions of Cell membrane in brief Protection, cell recognition, shape, storage, cell signalling.
   3.5 Cell Junctions: Tight junctions, gap junctions, Desmosomes.
4. **Nucleus: Structure and Function** **(04)**
   4.1 Introduction to Nucleus
   4.2 Structure of Nucleus: Nuclear envelope, Nuclear pore complex, Nucleoplasm, Nucleolus

4.3 Chromatin: Euchromatin and Hetrochromatin, nature and differences.

4.4 Functions of nucleus

5. Endomembrane System **(04)**

5.1 Introduction

5.2 Structure, location and Functions: Endoplasmic Reticulum, Golgi apparatus, Lysosomes and vacuoles.

6. Mitochondria and Peroxisomes **(03)**

6.1 Introduction

6.2 Mitochondria: ultrastructure and function of mitochondrion.

6.3 Peroxisomes

7. Cell Division **(05)**

7.1 Introduction

7.2 Cell cycle (G1, S, G2, M phases),

7.3 Mitosis.

7.4 Meiosis.

❖❖❖

Contents ...

❖❖❖

PATTERN OF UNIVERSITY EXAM QUESTION PAPER

Total Marks : 35 **Duration : 2 Hours**

Note :

(a) Q. 1 is compulsory.

(b) Solve any three questions from Q.2 to Q.5.

(c) Questions 2 to 5 carry equal marks.

1. Attempt any five of the following :

(a)

(b)

(c)

(d)

(e)

(f)

(Ask four tricky questions and two questions based on problem solving, if applicable.) (5)

2. (A) Describe – type of question(s) with internal options. (6)

(B) Short question, but tricky. (4)

3. (A) Explain – type of question(s) with internal options. (6)

(B) Problem based question, if applicable otherwise justification type – question. (4)

4. (A) Discuss - type of question(s) with internal options. (6)

(B) Problem based question, if applicable otherwise tricky and signified question. (4)

5. Write short notes on any four of the following : (10)

(A) Principle based.

(B) System based.

(C) Structure based.

(D) Descriptive based.

(E) Working based.

(F) Model based.

❖❖❖

1

CHAPTER

INTRODUCTION TO CELL BIOLOGY

CONTENTS

1.1 INTRODUCTION TO CELL BIOLOGY

The cell is the smallest unit of life. All living organisms are composed of single or many cells and cell products. The first compound microscope was invented by **Robert Hooke** and he first observed and described cells in 1665. He first used the term 'cell' to describe the hollow box-like cavities in cork. This was the beginning of a new branch of biology - the study of cell or **Cytology**. After this discovery, many other scientists made the observations of the cell and its various components in the next hundred years or so. Among them were the well known scientists, **Lamarck** (1809), **Dutrochet** (1824) and **Turpin** (1826). The botanist **Brown** (1831), discovered that all plant cells possess nucleus. In 1838, **Schleiden** a German botanist studied plant cells and discovered that all tissues of plants are made up of cells. He emphasised that cells are organisms and all animals and plants are aggregation of these organisms arranged according to certain basic laws. The German zoologist **Schwann** (1939) arrived at the same conclusion for animals. He stated that all the organisms are composed of essentially alike parts namely cells. This concept was then applied to all living organisms. The finding of **Schleiden** and **Schwann** that all living organisms consist of cells and cell products is now referred to as the cell theory or the cell principle.

According to the cell theory, all living organisms, from the simplest unicellular organisms to very complex plants and animals, is composed of cells and cell products and each cell can act independently, but functions as an integral part of the complete organism. Though this cell theory is adopted by both zoologists and botanists; there were objections to this theory in those days. This theory had some errors. For example, Schwann believed that cells could be spontaneously generated by a process analogous to crystal formation. But studies of developing embryos showed that during growth, cells multiply themselves by cell division. These observations were summarised by **Kollikar** (1841) that sperms and ova are histological elements originating in the organisms. **Rudolf Virchow** (1855) added a very important point in the cell theory with his famous aphorism (proverb) that all cells arise from pre-existing cells. An animal arises only from an animal and the plant only from a plant. This established cell division as central phenomena in the reproduction of organisms. Years later, it was shown that cells ensure continuity between one generation and another by the mechanism of mitosis (**Flemming,** 1880) during which the precise partitioning of chromosomes occur (**Waldeyer**, 1890).

Another important discovery was that the development of an embryo starts with the fusion of two nuclei, one coming from an egg and the other from a sperm cell that was introduced into the egg during fertilisation **(Hertwig**, 1875). Before the end of the 19th century, it was established that gametes (eggs and sperm cells) are formed by reduction of division, later called meiosis, by which the number of chromosomes of a species remains constant from one generation to another.

The Cell: The cell was first described in 1665 by **Robert Hooke**, the curator of Royal Society of London. While observing thin slices of cork under crude microscope, Hooke observed walls of cork cells which appeared in the form of small boxes or compartments. These compartments were later described as cells. (L. cella = compartment).

1.2 CELL AS BASIC UNIT OF LIFE

Although it is very difficult to define a cell, many scientists have defined it in their own way. Some of the definitions are as follows.

i) "Cell is the structural and functional unit of living organisms".

ii) "Cell is the smallest living unit capable of independent existence".

iii) "Cell is the smallest mass of living matter (protoplasm) containing nuclear material".

iv) "Cell is a piece of nucleated cytoplasm surrounded by cell wall or plasma membrane".

None of the above definitions is ideal as exceptions of each definition are available. **Loewy and Siekevitz (1963)** gave the definition of cell which has been accepted by majority of the scientists. The definition is as follows

"Cell is the unit of biological activity delimited by a semipermeable membrane and capable of self reproduction in a medium free from other living systems."

1.2.1 Divisions of Cell Biology

Cell biology is a hotspot in biological research facilitated by new and advanced instruments and techniques. For this purpose, the most significant and encouraging development in the 20th century has been the alliance of cell biology with the other fields of life sciences and interdisciplinary sciences. A brief account of the different branches of cytology are given below.

1. **Cytogenetics:** It is a branch of cytology which deals with the cytological and molecular basis of heredity, variation, mutation, phylogeny, morphogenesis, differentiation and evolution of chromosomal behaviour of organisms. For this, the study of cells, heredity and evolution was converged and was stated by Wilson as "Heredity appears as a consequence of the genetic continuity of the cells by division". The study of germ cells proved the "**Theory of continuity of germ cells**" proposed by **Weismann** (1833) to explain the transmission of heredity characters, due to the cell nucleus which contains the chromosomes as physical basis of heredity. The knowledge of cytogenetics, has been useful in the identification and synthesis of hereditary material, cracking the genetic code, hybridisation for the development of superior qualities of crop

varieties. Cytogenetics has also helped us to understand genetic and phenotypic expression, genetic control of growth and understanding the role of genes in the mechanism of organic evolution, etc.

Thus, cytogenetics helps in the study of various aspects of organisms on the basis of the behaviour of its nuclear material.

2. **Cytotaxonomy:** It is found that every species, whether plant or animal has a definite number of chromosomes in its cells, hence, it is similar to another organism of its kind in shape, size, habitat and behaviour. These characteristics of the chromosomes assist taxonomists in determining the taxonomical position of a species in taxonomy. It also helps in understanding the origin of certain taxonomic units, hence helps in determining the taxonomic position of any species.
3. **Cell Physiology:** Cell physiology deals with the study of life activities such as nutrition, metabolism, excitability, growth, reproduction (cell division) and cell differentiation. Thus cell physiology has helped in understanding various complicated physiological activities at the cellular level and this knowledge can be applied for understanding various normal and abnormal physiological life activities.
4. **Cytopathology:** The diseases that are caused due to abnormal activity of cells, tissues and toxic chemicals, for example, cancer cells, effects of viruses on cell structure and function, and effect of toxic chemical on tissues are studied with cytopathology. Thus it shows a close relation between cytology and pathology on one hand and medicine on the other hand; this in turn is useful in understanding various diseases of animals.
5. **Cytochemistry:** Cytochemistry is the branch of cell biology concerned with the study of physio-chemical and chemical composition of living matter. The cytochemical analysis indicates the presence of carbohydrates, proteins, lipids, nucleic acids and other organic and inorganic chemical components in the cells. With the help of biochemistry, researchers have been able to isolate and analyse cell molecules and organelles and their metabolic functions.
6. **Molecular Biology and Ultra structure:** This branch deals with the ultra structure of cell organelles, cell membranes and their molecular organisation with the help of electron microscopy. It is a modern branch of biology due to the merging of cytology with biochemistry,

physiochemistry, macromolecular and colloidal chemistry. Knowledge of ultra structure of the cell is of fundamental importance because practically all the functional and physiochemical transformations take place at the molecular level.

7. **Cyto-evolution:** Radiant energies such as ionising radiations (x-rays and γ-rays) and some chemicals such as CO^{60} (Cobalt 60) alter the structure and number of chromosomes. These alterations manifest themselves in the morphology and physiology of the organisms; it reveals to some extent how the evolution has occurred.
8. **Cyto-ecology:** It is the branch of cytology which deals with effects of ecological changes on the chromosome number of the cell. The cytological studies of plant and animal cells indicate the relation of the organism with its ecological habitat and its geographical distribution.
9. **Cell Culture:** Cell cultured *in-vitro* have been used for understanding basic cell biology, that is, DNA replication and problems thereof, transcription, protein synthesis, energy metabolism, receptor studies, cell cycle analysis, cell division, gene expression, signal transduction, apoptosis, effect of external agents like drugs and contaminants, effects of radiation, population kinetics, cell-cell and cell-matrix interactions and many other aspects. Cell cultures are also used to understand disease processes better, especially malignancy, isolation of viruses, vaccine production and production of biologically important molecules. Cell fusion techniques are used to understand somatic cell genetics, for mapping genes on chromosomes and for the development of monoclonal antibodies. Cell cultures are popularly used for gene transfer studies. Recently, stem cell cultures have been reported to differentiate into functional cells in-vitro, which have future applications in regenerative medicine. Thus, for all the above purposes, plant and animal cell cultures have been established in sophisticated research laboratories.

1.2.2 Shape and Size of Cells

The eukaryotic cells exhibit various forms and shapes. The animal cells are generally spherical or oval but cell may be cuboidal, polygonal, tubular, triangular, irregular or cylindrical. Depending on the nature or function, the shape of the cell also shows variation. Even the cells of the same organ may display variations in shape.

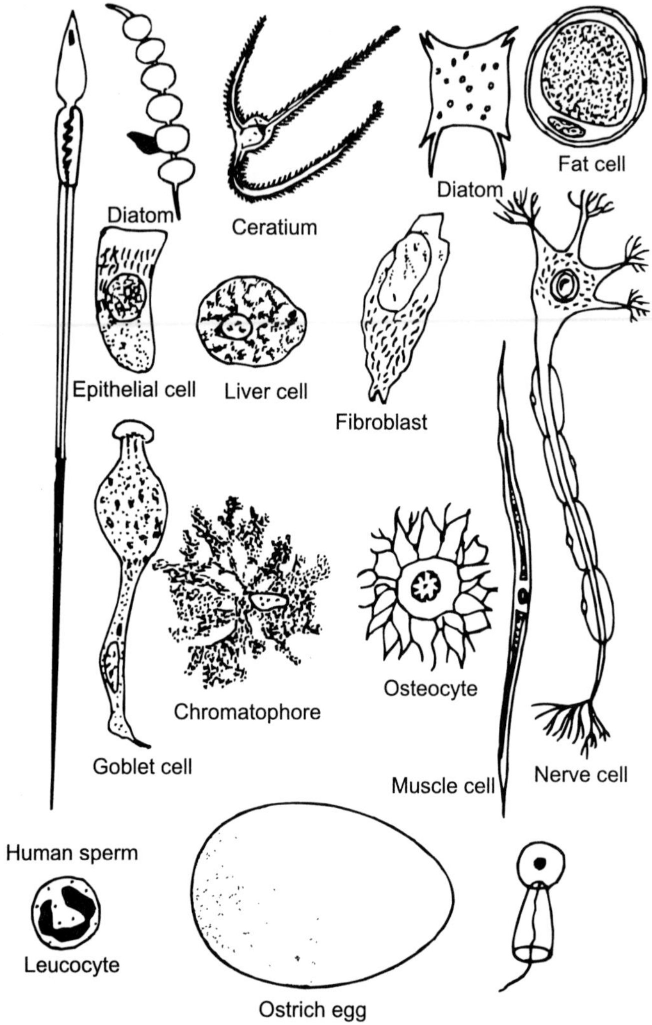

Fig. 1.1: Various types of eukaryotic cells showing different shapes

In case of animal, following different shapes are exhibited by the cells.

(1) Variable or Constantly changing shape: The protozoans like *Amoeba* and *Leucocytes* display variable shape.

(2) Fixed shape: Majority of the cells show fixed shape which remain constant throughout their life span

- **(i) Cuboidal:** The cells of squamous epithelium, endothelium and the upper layer of epidermis and the cells of thyroid gland follicles are cuboidal in shape.

(ii) Columnar: The cells lining the intestine are columnar in shape.

(iii) Discoidal: The erythrocytes or red blood cells of vertebrates are disc-shaped.

(iv) Spherical: The eggs of many animals are rounded or spherical in shape.

(v) Spindle shaped: The smooth muscle cells are elongated and spindle shaped.

(vi) Elongated: Nerve cells are elongated in shape.

(vii) Branched: The pigment or chromatophore cells of skin are branched.

The size of the cells vary from very small cells of bacteria (0.2 to 5.0 μ) to very large egg of the ostrich (175 mm in diameter). The Ostrich egg is considered as the largest cell. The longest nerve cells have been found to have a length of 0.92 to 1.06 metre. The following factors govern the size of the cell

(i) The nucleocytoplasmic ratio: It is the ratio between the volume of the nucleus and the cytoplasm.

(ii) The ratio of the cell surface to the cell volume.

(iii) The rate of metabolism.

1.3 IMPORTANCE OF CELL BIOLOGY AND ITS APPLICATIONS IN INDUSTRY

The cell is a fundamental, structural and functional unit of living organisms, just as the atom is the fundamental unit in chemical structures. The cell can be considered to be an organism in itself, and composed of many elements, the sum of which not only constitutes the cellular unit, but has particular significance to the organism as a whole. Our advanced knowledge of the cell has produced a fundamental change in the interpretation of cellular structures.

At a cellular level, the ultra structure is concerned with the larger repeating units that can be analysed with microscopic techniques. Electron microscopy has given us a peek into the various submicroscopic structures of a cell that were not known to exist before thus contributing greatly to our knowledge in the field of cytology.

The cell can be studied from various aspects such as morphology, biochemistry, physiology, genetics, development and molecular structures which help in proper understanding of various complicated life activities like metabolism, growth, differentiation, heredity and evolution at the cellular and molecular levels.

1.4 STRUCTURE OF PROKARYOTIC AND EUKARYOTIC (PLANT AND ANIMAL) CELL

Cell biology or cytology is one of the youngest branches of life sciences. The term 'cell' was first used by **Robert Hooke (1665)** to describe his investigations on "the texture of cork by means of magnifying lenses". Cell is the structural and functional unit of all the living beings.

Cytological studies present two main problems, the exceedingly small size and the transparent nature of cell and its components. Majority of cells are too small to be seen with the naked eye. It was only after the invention of microscope that the existence of cell was noticed. Since then the search continues for new instruments and techniques that can provide more and more details of the cell and its organelles. Microscope has opened a new micro-world which was invisible to us. Certain techniques have also been perfected during the course of time. These include the use of different types of stains or dyes, the process of centrifugation, cell-fractionation, chromatography, X-ray diffraction, autoradiography and tissue culture techniques, etc.

The term *cytology* (Gr. *kytos* - hollow) was used in 18th century for the microscopic study of the cell. It provides the detailed morphology of various cell organelles. This term has been replaced by cell biology, which includes the structural, genetical, physiological, biochemical and biophysical studies of cell and also attempts to explain the phenomenon of metabolism, biosynthesis, sex, variations, mutations and evolution of living organisms in terms of macromolecules (DNA, RNA and proteins).

1.4.1 Structure of Cells

The body of all living organisms (bacteria, blue green algae, plants and animals) except viruses have cellular organisation composed of single or many cells. The organisms which show only single cell in their body are called *unicellular organisms*. For example, bacteria, blue green algae, protozoa, etc. are unicellular. The organisms with many cells in their body are called *multicellular organisms.* Most plants and animals are multicellular. Thus, any cellular organism may contain either a

(I) Prokaryotic cell or

(II) Eukaryotic cell

1.5 STRUCTURE AND FUNCTIONS OF PROKARYOTIC CELLS

The prokaryotic (Gr., *Pro* = primitive; *Karyon* – nucleus) cells are most primitive and simple cells which occur in bacteria and blue green algae. These cells have primitive nuclei without nuclear envelope or membrane. The genetic material i.e. DNA, RNA and proteins etc. are in direct contact with cytoplasm. The prokaryotic cells lack cytoplasmic organelles like mitochondria, endoplasmic reticulum, golgi complex, lysosomes, centrioles, etc.

Bacteria

The bacteria are microscopic, single celled living organisms which do not possess chlorophyll. They are found in soil, air and water as well as in plants and animals. The cell structure is simpler than other organisms. They do not have a true nucleus or membrane bound organelles. Their genetic information is contained in a single loop of DNA present in the cytoplasm. They show different shapes like bacillus (rod), coccus (spherical), spirillum (spiral) and spirochete (corkscrew). Cellular arrangements occur either singularly, in chains, or in clusters.

Ultra Structure of Bacteria (*E. coli*)

A typical prokaryotic cell of bacteria has the following structures: cell wall and plasma membrane, a cell envelope consisting of a capsule, appendages (attachments to the cell surface) in the form of flagella and pili (or fimbriae), cytoplasmic region that contains the chromosome (DNA) and ribosomes; and various inclusions.

(i) Cell wall: The cell wall is made up of peptidoglycan complexed with teichoic acids. In some bacteria, it is surrounded by the outer membrane. The cell wall prevents osmotic lysis of cell protoplasm and gives shape and rigidity to the bacterial cell.

(ii) Plasma membrane: The cytoplasm of the bacterial cells is enveloped by a thin plasma membrane which is composed of lipid and protein molecules. The plasma membrane serves as a permeability barrier. It contains oxidative or respiratory chain enzymes and multi enzyme complexes which perform the function of mitochondria in eukaryotic cell. The plasma membrane show invaginations in cytoplasm called mesosomes. Mesosomes are useful as the respiratory organs and synthesise materials for extracellular transport and for formation of cell wall.

(iii) Capsule: The outer slimy layer of polysaccharides is called capsule. It is an additional outer covering that is present in some bacterial cells. It provides protection against phagocytic engulfment, helps the cell adhere to surfaces and nutrients and prevents desiccation by retaining the moisture.

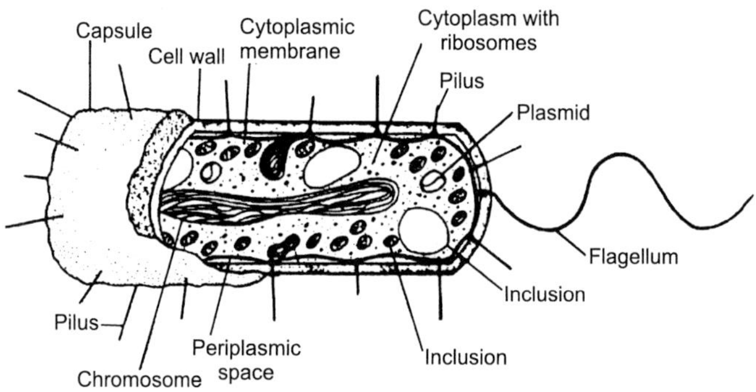

Fig. 1.2: A typical bacterial cell

(iv) Flagella: They are hair like appendages that protrude from the cell wall. Flagella are helically shaped structures containing the protein flagellin. It is not sheathed by the plasma membrane. The flagellum rotates in clockwise or anti-clockwise direction to propel the cell. Although the primary function of flagella is locomotion it is also used for attachment of bacterial cell to solid substratum. Flagella are much longer than the pili and are present on the bacterial cell in low numbers.

(v) Pili: They are hair-like appendages on the surface of the cell that attach to other bacterial cells. They are composed of helically arranged protein called pili. Pilus is required for bacterial conjugation. Shorter pili called fimbriae help bacteria attach to surfaces. They are generally shorter in length as compared to flagella and are present throughout the bacterial cell surface.

(v) Cytoplasm: The cytoplasm of a bacterial cell is dense, colloidal with granules of glycogen, proteins and fats. Except ribosomes; mitochondria, lysomes, endoplasmic reticulum and other organelles are absent in the prokaryotic cytoplasm. The ribosomes occur freely in cytoplasm and they form polyribosomes during protein synthesis. Certain photosynthetic bacteria show the presence of chromatophores having bacteriochlorophyll. In the central portion of the cytoplasm, occur one or two nucleus like clear zones. Each zone contains a single, large, circular, double stranded DNA molecule which is called bacterial chromosomes or genophores. These distinct nuclear regions are called nucleoids.

Escherichia coli

It is a gram-negative, non-pathogenic rod shaped bacterium commonly found in the lower intestine of warm-blooded organisms. It is commonly used as model organism by cell biologists because it divides rapidly and can be easily cultured in laboratory conditions .

E. coli is about 2 μm long and 0.8 to 1 μm thick and has outer cell wall of 100 Å thickness. E. coli possess flagella and are motile. The cell wall consists of an outer membrane containing lipopolysaccharides, a periplasmic space with a peptidoglycan layer, and an inner, cytoplasmic membrane. *E. coli* has only one circular chromosome which consists of double-stranded DNA molecule of 1400 μm length.

E. coli is commonly used as an indicator of feacal contamination of water bodies. This organism which is the normal flora of our intestine helps our body to metabolise the food, vitamin K production, and food absorption.

1.6 STRUCTURE AND FUNCTIONS OF EUKARYOTIC CELLS

The eukaryotic cells (Gr. eu = good, karyotic = nucleated) possess true nucleus, i.e. the nuclear material such as DNA remain well separated from the cytoplasm by two porous lipoproteinous nuclear membranes. The eukaryotic cells are easily distinguished from a prokaryotic cell through a membrane-bound nucleus. Eukaryotic cells also contain many internal membrane-bound structures which perform various metabolic functions. The eukaryotic cells occur in most plants (from algae to angiosperms) and animals (from protozoa to mammalia). These cells are complex in structure and show different shapes and sizes.

These cells are larger than prokaryotic cells. The size of the cells varies from 1 μ to 175 mm; for example, ostrich egg is the largest cell with 175 mm in diameter. Nerve cells are the longest cells having a length of 3 - 3.5 feet.

Cell wall may or may not be present. The cytoskeleton is composed of microfilaments, intermediate filaments and microtubules. The eukaryotic DNA is organised in one or more linear molecules called chromosomes. The chromosomes are always associated with histone proteins. The chromosomal DNA is stored in the cell nucleus. Some eukaryotic organelles like mitochondria and chloroplasts also contain DNA.

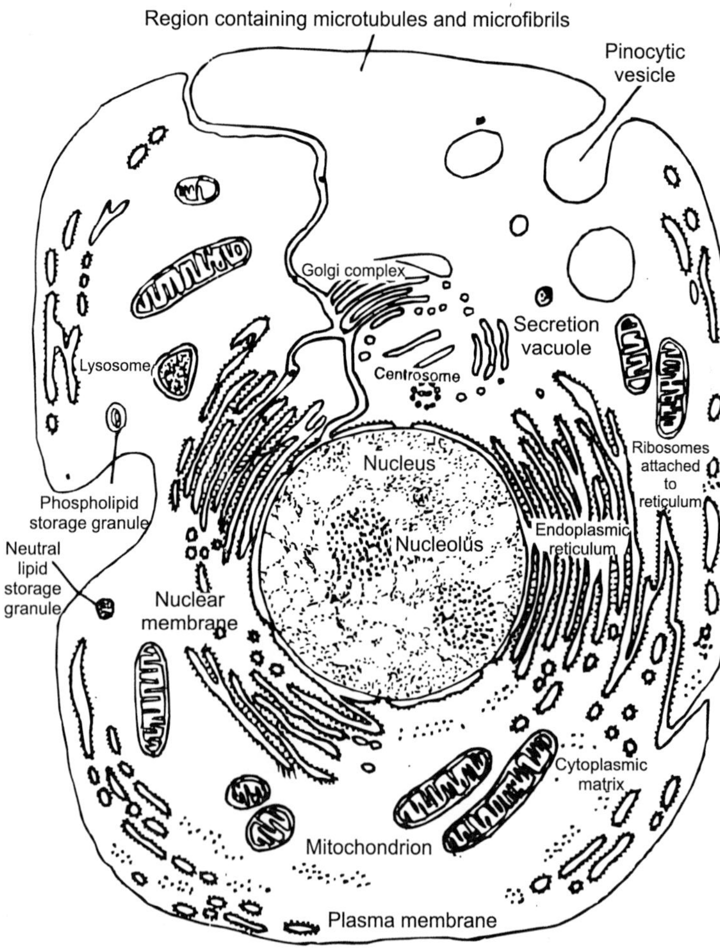

Fig. 1.3: A typical eukaryotic animal cell

1.6.1 Structure of Eukaryotic Plant Cell

The plant body consists of a number of small microscopic box like compartment called the cell. Robert Hooke was the first to discover plant cell by studying a piece of cork under the microscope. He saw a honeycomb like structure in it, and each compartment was referred to as a cell (Latin: cellular-small compartment).

The plant cells contain structures such as a cell wall, plastids, and large vacuoles. As the plant matures, its cells become specialised in order to perform certain functions necessary for survival. The cells are variable in size and shape. The shape may be spherical, oval, polygonal, rectangular or elongated. The average size of fully developed cell varies between $\frac{1}{10}$ millimeter to $\frac{1}{100}$ millimeter.

A typical plant cell has an outer thick covering called the cell wall which encloses a colourless semifluid granular substance called the protoplasm. The protoplasm is the fundamental substance of all organisms.

The following structures and organelles are present in a typical plant cell.

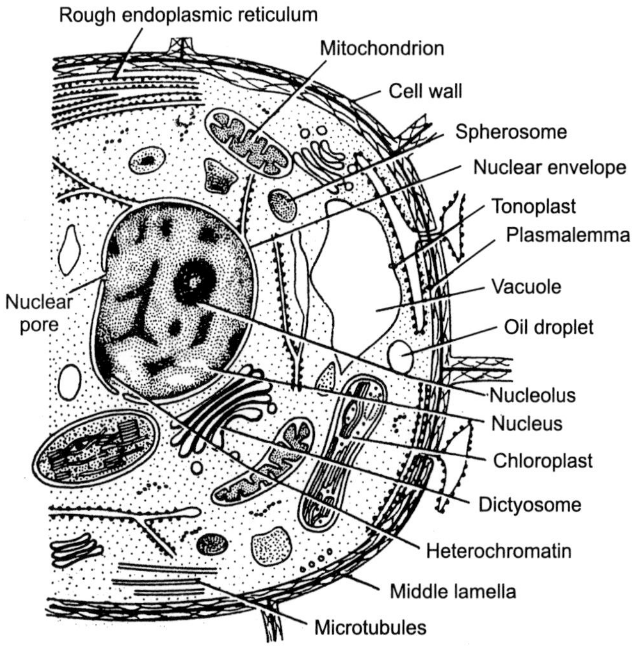

Fig. 1.4: Detailed structure of plant cell

1. Cell Wall

It is a rigid and elastic membrane bounding the cell. The plant cell wall is made of cellulose $[(C_6H_{12}O_5)n]$ which is the most abundant macromolecule on earth. The cell wall possesses a large number of minute pores called micellae, through which different cells are interconnected. The cell wall acts as a semipermeable membrane.

The plant cell walls consist of three layers:

- **Middle lamella:** It is the outer wall of the cell and is shared by adjacent cells. It is composed of pectic and protein.
- **Primary wall:** It consists of a rigid skeleton of cellulose microfibrils embedded in a gel-like matrix composed of pectic compounds, hemicellulose and glycoproteins.
- **Secondary wall:** It is formed after the cell is enlarged. It is made of cellulose, hemicellulose and lignin. The secondary wall is generally layered.

2. Cell (Plasma) Membrane

It is a thin, semi-permeable membrane that surrounds the cytoplasm of a cell, enclosing its contents. It lies just below the cell wall.

3. Cytoplasm

It is a gel-like, transparent, semifluid substance present in the cell in which the nucleus and different organelles are embedded. Cytoplasm is the site where most cellular activities occur, such as many metabolic pathways and cell division. In plants, movement of the cytoplasm around vacuoles is known as cytoplasmic streaming.

4. Cytoskeleton

It is a network of fibres throughout the cytoplasm that helps the cell maintain its shape and gives support to the cell. The cytoplasm comprises of

- **Microfilaments:** These are solid rods made of globular proteins called actin. These filaments are primarily structural in function and are an important component of the cytoskeleton.
- **Microtubules:** These are straight, hollow cylinders found throughout the cytoplasm of all eukaryotic cells. They are absent in prokaryotes. They carry out a variety of functions, ranging from transport to structural support.

5. Nucleus

Nucleus is a spherical, highly specialised, dense part of the protoplasm embedded in the cytoplasm. Although all vital activities in the cell take place in cytoplasm but they are controlled by the nucleus. Thus, nucleus is the control center of the cell. The nucleus is surrounded by nuclear membrane. It serves two major functions

- It stores the cell's genetic material.
- It co-ordinates the cell's activities, which include growth, metabolism, protein synthesis, and cell division.

On dry weight basis, the following chemical substances are found in the nucleus

Proteins	70 %
Phospholipids	3 – 5 %
DNA	10 %
RNA	2 – 3 %

The nucleus comprises of

a. Nuclear membrane = Nuclear envelope = *Karyotheca*

b. Nuclear sap = Nucleoplasm = Karyolymph

c. Nucleolus

d. Chromatin threads

The nuclear membrane is a lipid bilayer which surrounds the genetic material and nucleolus. The nucleolus within the nucleus helps in the synthesis of ribosomes. Nuclear pores are tiny hole within the nuclear membrane that allows nucleic acids and proteins to move into and out of the nucleus. The most conspicuous nuclear components are the slender threads of chromatin (DNA, RNA and nucleoproteins.) During cell division these thicken, and each has a distinct shape. They are then known as chromosomes. The number of chromosomes for each cell nucleus is constant for a given organism. This is known as the chromosome number.

6. Chloroplasts

They are specialised organelles that carry out the function of photosynthesis. Chloroplasts are made up of stacks of thylakoid discs that contain chlorophyll. A stack of thylakoid discs is called a granum. Photosynthesis (the production of ATP molecules from sunlight) takes place on thylakoid discs.

7. Endoplasmic Reticulum (ER)

The endoplasmic reticulum is a network of sacs that manufactures, processes, and transports proteins in the cell. The rough ER is a vast system of interconnected, membranous, infolded and convoluted sacs located in the cell's cytoplasm. The ER is continuous with the outer nuclear membrane. Rough ER is covered with ribosomes that give it a rough appearance. Rough ER transport materials through the cell and produces proteins in sacs called cisternae. The smooth ER buds off from rough ER, moving the newly-made proteins and lipids to the golgi body and membranes

8. Golgi Complex

It is responsible for manufacturing, storing and transporting certain cellular products. It modifies proteins and fats made in the endoplasmic reticulum and prepares them for export outside the cell.

9. Mitochondria

They are oblong shaped, double membrane organelles found in the cytoplasm. The inner membrane is folded to form a series of projections called cristae. The mitochondrion converts the energy stored in glucose into ATP (adenosine triphosphate) for the cell.

10. Peroxisomes

These are tiny structures bound by a single membrane that contain enzymes that produce hydrogen peroxide as a by-product. These structures are involved in photorespiration.

11. Plasmodesmata

These are pores or channels between plant cell walls that allow molecules and communication signals to pass between individual plant cells.

12. Ribosomes

They are tiny organelles composed of approximately 60 % RNA and 40 % protein. They are the sites of protein synthesis.

13. Vacuoles and Cell Sap

The cytoplasm of mature cells contain one or more cavities, whose composition may vary in different cells and these cavities are termed as vacuole. These cavities or vacuoles are filled with cell sap. The cell sap contains various salts, glucose, sugar, albuminoids and organic acids. Sometimes vacuoles may contain oils, essences, resins etc. The vacuoles are of great significance for storage, detoxification, protection, and growth of a plant cell.

1.6.2 Ultra Structure of Animal Cell

A typical eukaryotic animal cell consists of the following:

1. Plasma membrane
2. Cytoplasm
3. Nucleus

1. Plasma membrane

The plasma membrane is trilaminar membrane of lipid and protein molecules which forms the external covering. It is ultra thin, elastic, porous, semipermeable membrane of the cell. It primarily provides mechanical support and shape to the cell. The plasma membrane also delimits the protoplasm from exterior, checks the entry or exit of the substances. Due to its semipermeable nature it allows the transport of only necessary materials across the cell. The ultra structure shows that the outer and inner layer of the plasma membrane is made up of proteins and middle layer is made up of lipids. The plasma membrane performs various important physiological functions such as osmosis, diffusion or passive transport, active transport, endocytosis and exocytosis etc.

2. Cytoplasm

The cytoplasm is distinguished into the following components:

(A) Cytoplasmic Matrix:

The portion between the plasma membrane and nucleus is filled by an amorphous, translucent, homogenous liquid known as cytoplasmic matrix or hyaloplasm. It is a complex mixture of various inorganic elements, salts and many organic molecules. The peripheral part is called *ectoplasm* and inner portion which is granular and less viscous is called *endoplasm*. The organic molecules present are carbohydrates, lipids, proteins, nucleic acid (DNA, RNA), nucleoproteins, nucleotides, amino acids, fatty acids and variety of enzymes.

(B) Cytoplasmic Structures:

The cytoplasmic organelles/ structures present in the cytoplasmic matrix are:

(i) Cytoplasmic inclusions: In the cytoplasmic matrix, many stored food granules and secretory substances of cell remain suspended in the form of refractile granules and form cytoplasmic inclusions. The cytoplasmic inclusions include oil drops, yolk granules, pigments, secretory granules and glycogen granules.

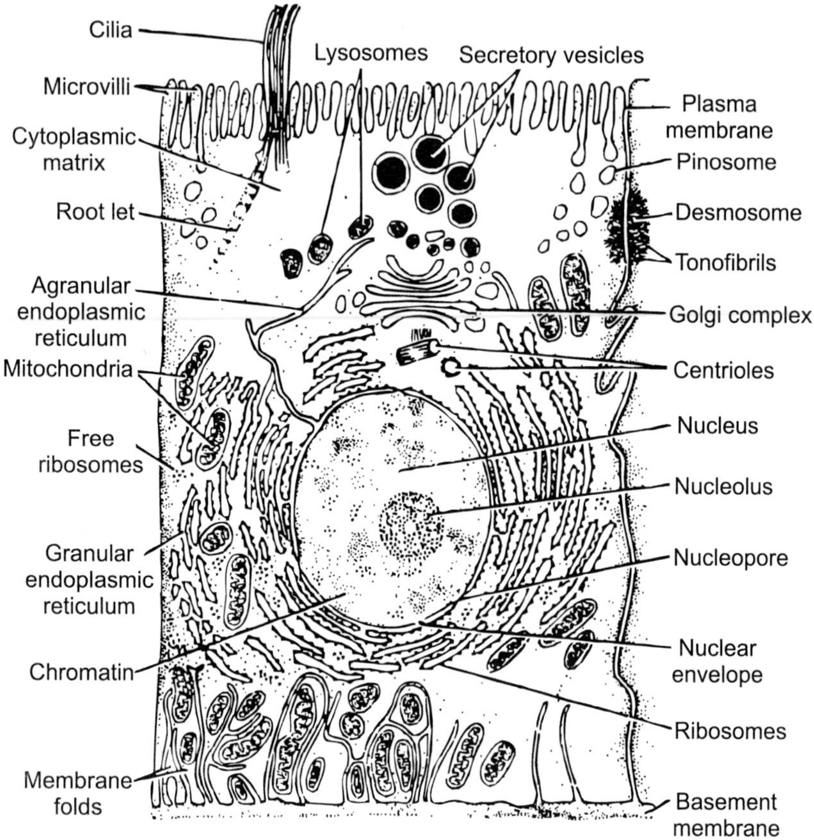

Fig. 1.5: Structure of a typical eukaryotic animal cell

(ii) Cytoplasmic organelles: The most important cytoplasmic organelles are endoplasmic reticulum, ribosomes, golgi complex, **lysosomes**, mitochondria and centrioles. They perform various important biosynthetic and metabolic functions such as storage, synthesis, transportation, support and reproduction.

(a) Endoplasmic reticulum: The endoplasmic reticulum forms the network of inter connecting tubules and vesicles. After each cell division, it forms a nuclear envelope around the nucleus and new plasma membrane between the dividing cells. It bears dense, granular and rounded bodies called ribosomes.

(b) Ribosomes: They are minute spherical structures attached to the endoplasmic reticulum. They are also found scattered freely in the cytoplasm. Ribosomes originate in the nucleolus and consist mainly of ribonucleic acid (r-RNA) and proteins. They are sites of protein synthesis.

(c) Golgi complex: It is a stack of flattened, membrane bound, parallely arranged organelle which occurs near the nucleus around centrioles and has connection with endoplasmic recticulum. Its main function is storage of synthetic products i.e. proteins of ribosomes and endoplasmic reticulum and secretion of certain macromolecules. They also produce secretory granules e.g. acrosome, a digestive enzyme present in the cap-like structure of sperms.

(d) Lysosomes: These are small, tiny; spheroid membrane bound vesicles which contain enzymes for intracellular digestion of foreign substances (bacteria etc.) and sometimes for the cellular organelles of the same cell.

(e) Mitochondria: They are rod shaped double membrane bound structures having the enzymes for respiratory metabolism. Mitochondria have their own circular DNA and ribosomes. They produce energy rich compound called adenosine triphosphate (ATP). Thus, they are called the power house of the cell.

(f) Centrosomes: They contain dense cytoplasm and are located near nucleus. During cell division centrosome is found to contain two rod-shaped granules called centrioles. At the time of cell division the centrioles form the spindle structure of microtubules which help in the separation and movement of chromosomes.

(g) Cytoplasmic vacuoles: These are small, liquid filled, structures useful for storage, transmission of materials and maintenance of internal pressure of the cell. They are found in both plant and animal cells.

(h) Microtubules: The plant and animal cells contain many fine tubules called microtubules in the cytoplasm for transportation of water, ions, small molecules and formation of fibres or asters of the spindle during cell division.

(i) Basal granules: The animal or plant cells having locomotory organelles such as flagella or cilia contain spherical bodies called basal granules or kinetosomes at the base of flagella or cilia.

3. Nucleus

It is usually a spheroid, centrally located cellular component which controls all the vital activities of the cytoplasm such as growth regulation and reproduction of the cell. It carries the genetic material of the cell. The nucleus consists of the following structures:

(a) **Nuclear membrane or envelope:** It is a perforated envelope which is made up of two lipoproteinous membranes. This membrane provides selective continuity between nuclear and cytoplasmic materials. It is continuous with the endoplasmic reticulum.

(b) **Nucleoplasm and chromatin fibres:** The nucleus is filled with colloidal ground substance called *nucleoplasm*. The nucleoplasm contains threadlike elongated structures known as *chromosomes*. The chromosomes appear only during cell division. Otherwise, they occur in the form of ultra thin chromatin threads.

(c) **Nucleolus:** It is conspicuous, darkly stained, intranuclear spherical body which is accentric in position. Chemically, it contains large amount of ribosomal proteins and ribosomal RNA. It plays an important role in the process of protein synthesis and cell division. Its main function is biogenesis of ribosomes.

Table 1.1: Distinguishing characters between prokaryotic cell and eukaryotic cell

S. N.	Character	Prokaryotic cell	Eukaryotic cell
1.	Nuclear	Absent	Present
2.	Mitosis	Does not occur	Occurs
3.	DNA arrangement	Circular	Linear
4.	Histones	Absent	Present
5.	Respiratory system	Part of plasma membrane, mitochondrion absent	In mitochondria

Contd...

6.	Photosynthetic apparatus	In internal membranes, chloroplasts absent	In chloroplasts
7.	Golgi bodies, Endoplasmic reticulum, lysosome	Absent	Present
8	Ribosomes	70S type	80S type
9.	Cell wall	Chemically complex	Only in plant cells and fungi, chemically simpler
10.	Flagella	Composed of only one fibre	Each flagellum composed of fibrils in a distinct 9 + 2 pattern.
11.	Cytoplasmic movements	Cytoplasmic streaming rare or absent	Cytoplasmic streaming often occurs
12.	Vacuoles	Absent	Present
13.	Lysosomes	Absent	Present in animals
14.	Capsule	May be present	Absent
15.	Size	1-10 μm	10-100 μm

POINTS TO REMEMBER

- The cell is a fundamental, structural and functional unit of living organisms.
- The different branches of cytology are cytogenetics, cytotaxonomy, cell physiology, cytopathology, cytochemistry, molecular biology and ultra structure, cyto-evolution, cyto-ecology, cell culture.
- The prokaryotes are the most primitive organisms.
- Prokaryotic organisms do not have a true nucleus.
- Eukaryotic cells possess a membrane-bound nucleus.

- Bacteria are microscopic, single celled, living organisms which do not possess chlorophyll.
- The bacterial cell wall is made up of peptidoglycan.
- The cell wall prevents osmotic lysis of cell and gives shape and rigidity to the bacterial cell.
- The plasma membrane is composed of lipid and protein molecules and it serves as a permeability barrier.
- Capsule is a slimy layer of polysaccharides which provides protection against phagocytic engulfment of the bacterial cell.
- Flagella are hair like appendages that protrude from the cell wall.
- Pili are hair-like appendages on the surface of the cell which are made up of helically arranged protein called pilin.
- *E. coli* is a gram-negative, non-pathogenic, rod shaped bacterium commonly found in the lower intestine of warm-blooded organisms.
- The plant cells contain structures such as a cell wall, plastids, and large vacuoles.
- The plant cell wall is chiefly made of cellulose.
- Chloroplasts are specialised organelles that carry out the function of photosynthesis.
- Plant cells have plasmodesmata which are pores or channels between plant cell walls that allow molecules to pass between individual plant cells.
- Eukaryotic cells exhibit various forms and shapes.
- The factors that affect the size of the cell are – the nucleocytoplasmic ratio, ratio of the cell surface to the cell volume and the rate of metabolism.
- The plasma membrane of an animal cell is a trilaminar membrane of lipid and protein molecules.
- The endoplasmic reticulum forms the network of inter connecting tubules and vesicles.
- The nucleus of a eukaryotic cell is a spheroid, centrally located cellular component which controls all the vital activities such as growth regulation and reproduction of the cell.

EXERCISE

1. Define the cell. Describe the various branches of cell biology.
2. Write short notes / Define / Explain the following.
   a. Branches of cell biology
   b. Cytogenetics.
   c. Cytotaxonomy
   d. Cell physiology
   e. Cytopathology
   f. Cyto-chemistry
   g. Molecular biology and ultrastructure
   h. Cyto-evolution
   i. Cyto-ecology
   j. Cell culture
3. Define / Explain the following terms.
   a. Cell
   b. Cytogenetics
   c. Cytotaxonomy
   d. Cell physiology
   e. Cytopathology
   f. Cyto-chemistry
   g. Cyto-evolution
   h. Cyto-ecology
   i. Cell-culture
4. Describe the ultra structure of a prokaryotic cell.
5. Describe the structure of a typical bacterial cell.
6. Draw a labelled diagram of prokaryotic cell.
7. Draw a labelled diagram of eukaryotic cell.
8. Give an account of the structure of an animal cell and state the functions which are attributed to its component parts.
9. Draw a labelled diagram of an animal cell.
10. Give the distinguishing features of prokaryotic and eukaryotic cell.

11. Differentiate between plant and animal cells.
12. What is cell? Give an account of different shapes and sizes of cells.
13. Describe the structure and functions of animal cell. **(April 2014)**
14. Write short notes on
    a. Prokaryotic cell
    b. Eukaryotic cell
    c. Nucleus
    d. Shape and size of cells.
15. Define / Explain.
    a. Prokaryotic cell **(April 2014)**
    b. Eukaryotic cell
    c. Cell

❖❖❖

2

CHAPTER

TECHNIQUES IN CELL BIOLOGY

CONTENTS

2.1 MICROSCOPY

Introduction

The word microscope is derived from two Greek words, *'micro'* = small and *'scope'* = to view. Thus, microscope is an instrument designed for visual examination of small object which can not be examined by the naked eye. There are different types of microscopes such as simple, compound, phase – contrast and electron microscope. The study of different microscopes is called *microscopy.*

The earliest recorded use of magnifying lens goes back to Conrad Gesber (1558), a Swiss biologist who published work on the structures of *foraminiferans* (Protozoans). It was Zacharias Janssen, a Dutch Astronomer, who in 1590 added another lens to the telescope and thus provided for the first time a prototype of the present day telescope and the compound microscope.

Anton Van Leeuwenhoek (1674), the father of biology, was the first to use the microscope for biological studies. The compound microscope was constructed by Robert Hooke (1665) and the forerunner of the present day compound microscope.

2.1.1 Simple Microscope

The simple microscope is commonly used in the laboratory which is nothing but a magnifying glass. It is used when very small animal have to be dissected, or for magnification of the parts in the sections or of small organisms which are not seen by naked eye. The simple microscope is made up of two parts (1) Viewing or optical part and (2) Mechanical part.

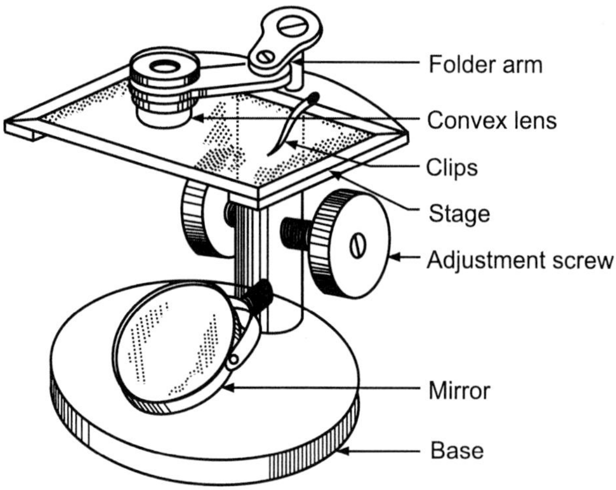

Fig. 2.1: Simple Microscope

Optical part consists of single convex lens. Whereas the mechanical part hold the lens in the proper position. With the help of knob proper magnification can be obtained. Mirror is present for light and clips for holding the slide.

A simple microscope consists of single lens or a combination of lenses which functions as convex lens. A convex lens magnifies the object and helps to produce a magnified image of a near object which appears to be at the distance of distinct vision. The enlarged image is formed on the retina. It shows details which cannot be seen by naked eye.

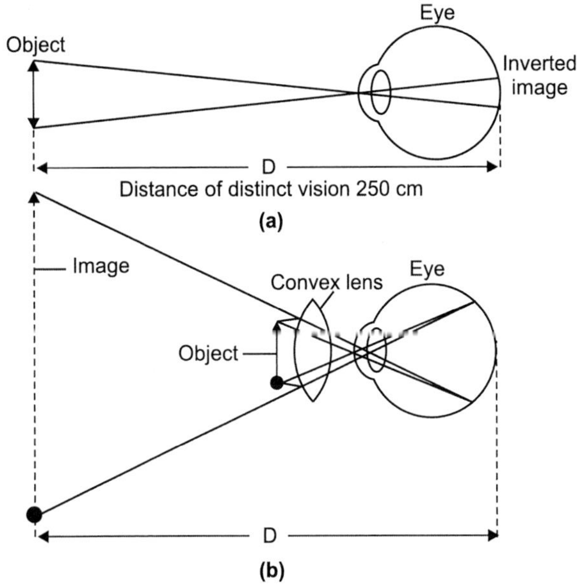

Fig. 2.2: (a) Perception of an Object by the Human eye at a Distance of Distinct Vision, (b) Perception of an Object through a Simple Convex Lens. Which Produce a Magnified Image at the Distance of Distinct Vision

The magnification obtained with a convex lens can be easily calculated by the following formula.

$$M = \frac{250 + 1}{f}$$

Where, f is the focal length of the lens in centimetres and 250 is distance of the distinct vision in centimetres. The magnification that can be obtained with a simple lens cannot exceed three times (3 × magnification). However, magnification more than 3 × 6 can be achieved by using a combination of several lenses. The combination of several lenses fused together acts as a single convex lens. The lenses fused together are called elements. Each element cancels out the distortions and defects produced by the other element. The net result of the combination is the formation of an image free from distortions. A magnification of about 20 times can be obtained. Such microscopes are used by biologists for smaller magnification during field work. It consists of a combination of double concave lens of crown glass fitted between two double convex lenses of flint glass. This is called an achromatic triplet or a planner lens. The three lenses are one cemented together and function as one lens.

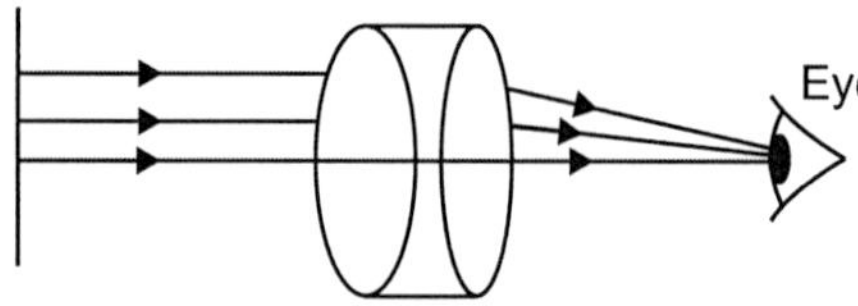

Fig. 2.3: An Achromatic Triple or a Planner Lens Showing the Path of Light Rays while Functioning as a Magnifier

2.1.2 Compound Microscope

A microscope magnifies the image of an object. The modern compound microscope which is also called light microscope, is most important optical instrument for the study of cells and tissues. It is important apparatus in a medical laboratory. It is a precision instrument and needs careful handling. Improper use of the microscope leads to the loss of clarity of the image.

The light microscope uses white light, either the external sunlight or the internal tungsten filament lamp, as a source of illumination. While observing under microscope, the objects either look dark or coloured, contrasted against a lighted background.

The Microscope and its Accessories:

Accessories or components of commonly used light monocular microscope is shown in Fig. 2.11. All the components of monocular and binocular microscopes are the same except that the binocular microscope has two eye pieces (ocular), which allow the user to keep both eyes open while viewing through the microscope. Both the microscope has three main systems such as support system, illumination system and magnification system.

I. Support System: It consists of the following parts:

(i) Foot Rest: It is used to hold the microscope while transporting.

(ii) Tube: It holds the optical system for magnification. It may be straight and fixed (older models), or bent and movable (modern models) for convenient viewing. The tube has eye piece at upper end i.e. near the eye. In case of bent tube, a special prism is put in the middle of the tube where it is bent. The prism directs the light beam coming through the objective lenses towards the eye piece. In binocular microscope, the light beam further shifts and reaches both the eye pieces.

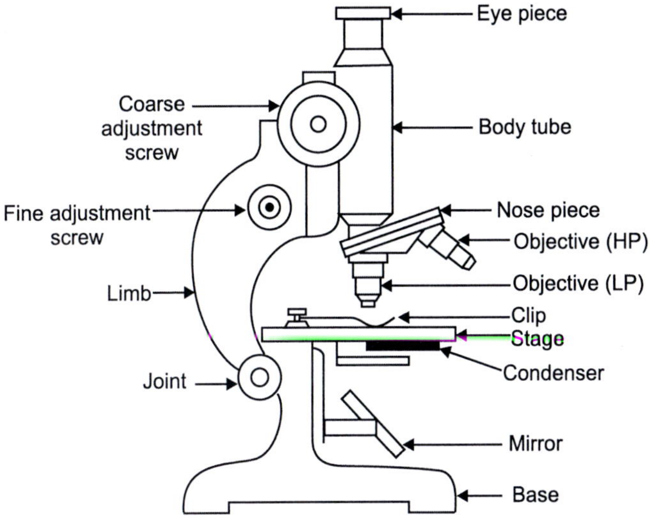

Fig. 2.4: Microscope

(iii) Stage: It is used to hold the object slide in place for viewing. The stage has central hole for passing the light in order to illuminate the object. The stage can be fixed or a mechanical stage can be screwed on to the fixed stage which can move the object slide across the stage or along the stage, horizontally or vertically. Focusing of optical system can be done either by moving the stage up and down (older models) or by the movement of the limb (modern models). The support system helps to carry the microscope in an upright position with one hand holding the base.

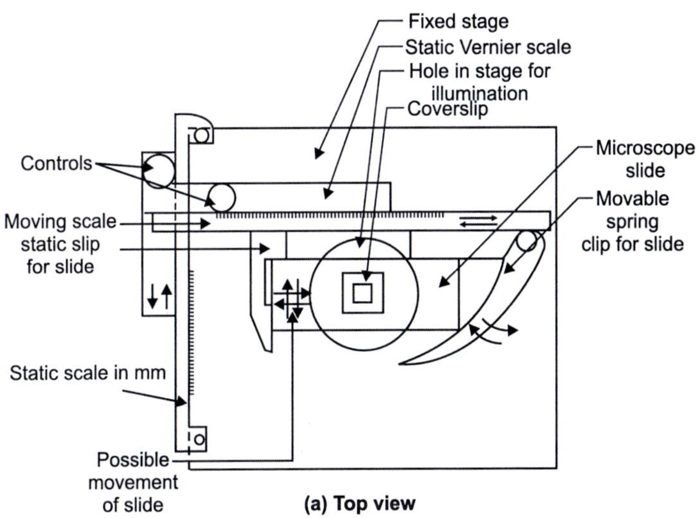

(a) Top view

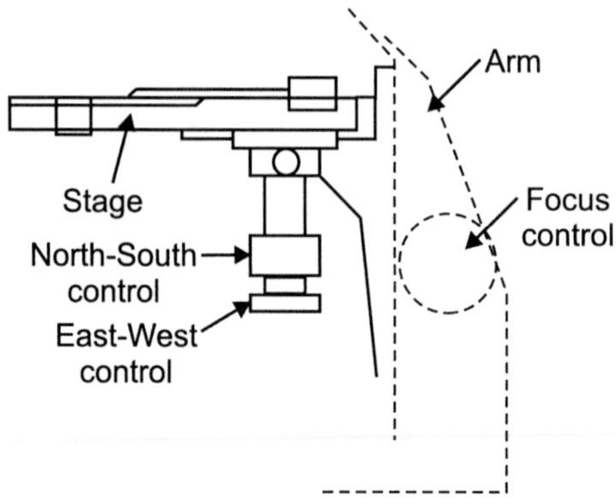

(b) Side view

Fig. 2.5: Parts of the Mechanical Stage from (a) Top View, and (b) Side View

II. Illumination System:

This system provides proper illumination to the object. It consists of the following parts:

(i) The Source of Light: Electric light should be preferably be used, since it is easier to adjust. It is provided by the lamp, placed in front of the microscope. An opaque electric bulb of the daylight type of 60 watts is used. Otherwise, daylight can be used.

(ii) Mirror: The mirror reflect rays from light source on the object. One side has a plane surface and the other side has a concave surface. The concave side forms a low power condenser and is used if there is no condenser.

(iii) The Condenser: The condenser brings the rays of light to a common focus on the object to examine. It is situated between mirror and the stage. It can be raised (maximum illumination) and lowered (minimum illumination). The common type of condenser consists of two or three lenses.

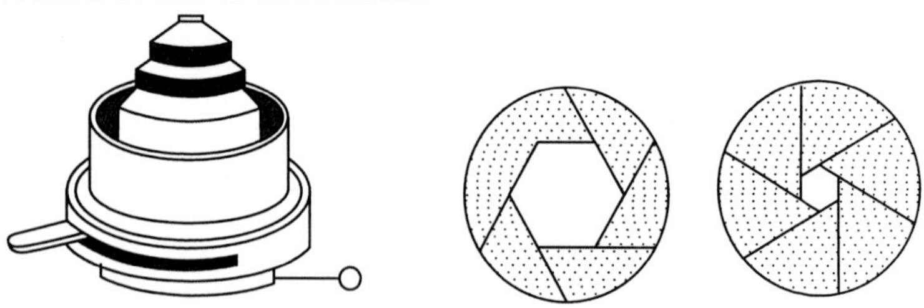

(a) Condenser (b) Diaphragm

Fig. 2.6: Parts of Illumination System

(iv) The Diaphragm: The diaphragm which is lodged below the condenser is used to reduce or to increase the angle and therefore also the amount of light that passes into the condenser. The upper lens of the condenser coincides in position with the aperture on the stage.

III. The Magnification System:

This consists of a system of lenses. The lenses of the microscope are mounted in two groups, one at each end of the long tube - the body tube. The first group of lenses is at the bottom of the tube, just above the preparation under examination (the object) and is called the *objective*. The second group of lenses is at the top of the tube where the microscope applies his eye and is called the eyepiece.

The Objective: The magnifying power of each objective is shown by a figure engraved on the sleeve of the lens.

×10 objective magnifies	10 times (low power)
×40 objective magnifies	40 times (high power)
×100 objective magnifies	100 times (oil immersion)

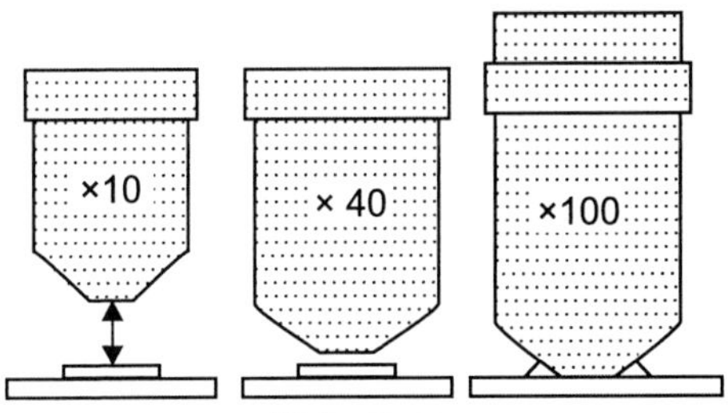

Fig. 2.7: Objectives of the Compound Microscope

The ×100 objective is usually marked with a red or black ring to show that it must be used with immersion oil.

The greater magnifying power of the objective, the smaller working distance i.e. distance between the front lens of the objective and the object slide when the image is in focus.

The eyepiece: The magnifying power of the eyepiece is marked on it.

- $\times$ 5 eyepiece magnifies the image 5 times
- $\times$ 10 eyepiece magnifies the image 10 times

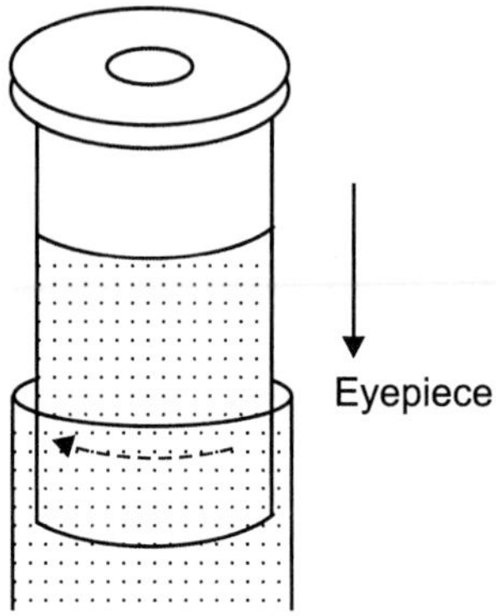

Fig. 2.8: Eyepiece

If you see an ×40 objective with an ×5 eyepiece, the object will be magnified 40 × 5 times, i.e. the total magnification will be 200.

IV. The Adjustment System:

The system comprise of the following:

1. **Coarse Adjustment Screw:** This is the largest screw. It is used first to achieve an approximate focus.
2. **The Fine Adjustment Screw:** This screw moves the objective more slowly. It is used to bring the object into perfect focus.
3. **The Condenser Adjustment Screw:** This is used to raise the condenser for greater illumination or to lower it to reduce the illumination.
4. **The Iris Diaphragm Lever:** This is small lever, fixed on the condenser. It can be moved to close or open the diaphragm, thus reducing or increasing both the angle and the intensity of the light.
5. **Mechanical Stage Controls:** These are used to move the object slide on the stage. One screw to move it backward and forward; and one screw to move it to left or right.

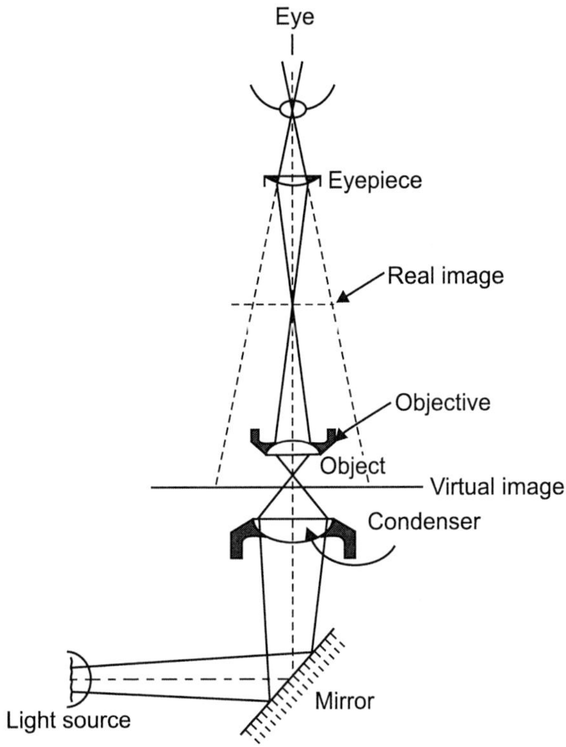

Fig. 2.9: The Path of the Light Waves to Form the Real Image Formed by the Objective. The Eye Sees the Magnified Virtual Image Formed

How to use Microscope ?

Following points should be remembered while using the microscope:

1. Place the microscope on a firm, level bench of adequate size but not too high.
2. Screw the objective into the revolving nose – piece following this order in a clockwise direction.

   × 10 objective.

   × 40 objective.

   × 100 objective.
3. Place the eyepiece in the body tube.
4. Fix the condenser under the stage.
5. Fix the mirror on the foot.

6. If electric illumination is to be used, place the lamp 20 cm in front of the microscope facing the mirror, which should to be fixed at an angle of 45°. Adjust the position of lamp so that it shines on the center of the mirror.
7. Use the plane side of the mirror. Open the iris diaphragm to the maximum. Raise the condenser. Place piece of thin white paper over the lens at the top of the condenser. This piece of paper should show an image of the electric bulb, surrounded by a circle of light. Adjust the mirror so that the image of the bulb is in the exact center of the circle of light (or the brightest part if daylight is being used).
8. It is very important to entire the condenser correctly.
9. Use the low power objective.

   Lower the condenser down to the bottom. Lower objective until it is just above the side preparation. Raise the objective, using the coarse adjustment screw, until a clear image is seen in the eyepiece. Occasionally a clear image can not be obtained although the objective has been lowered as far as possible. This is because the fine adjustment screw has been termed right to the end. Turn it back as far as it will go in the other direction and then focus by raising the objective.
10. Using high power objective (×40): Lower the condenser half way down. Lower the objective until it is just above the slide preparation. Using coarse adjustment, raise the objective very slowly until blurred image appears on the field. Bring into focus, using the fine adjustment. Raise the condenser to obtained sufficient illumination.
11. Using the oil immersion objective (×100): Perfectly dry stained preparation must be used. Place a tiny drop of immersion oil on the part to be examined (use synthetic oil, which do not dry, in preference to cedar – wood oil, which dries quickly).

    Raise the condenser up as far as it will go and open the iris diaphragm fully. Lower the ×100 objective until it is in contact with the oil. Bring it as close as possible to the slide, but avoid pressing on the preparation. Look through the eye piece and turn the fine adjustment very slowly upwards until the image is in focus. If the illumination is inadequate, use the concave side of the mirror.

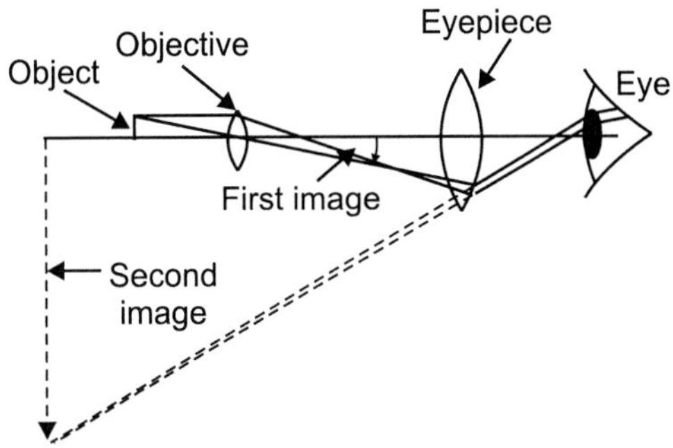

Fig. 2.10: Image Formation in a Light Microscope Occurs in Two Stages

12. Remember that the circle of light seen in the eyepiece is called microscope field.
13. Objective seen at the bottom of the field are actually at the top. Objective seen on the left hand side of the field are actually on the right. If you move the slide to the right, the object examined moves to the left. If you move the slide towards you, the object examined moves away from you.

2.1.3 Electron Microscope

The Electron Microscope (EM) is an optical instrument which utilizes electrons as a source of illumination for observing very minute object as a great magnification. It is a product of a large number of inventions and discoveries in physics. EM can achieve a very high power of resolution because it uses electrons of much shorter wavelength. Today there are several types of electron microscopes available to cell biologists.

Both the light and electron microscopes are similar in arrangement and function of their components and consists of three major systems

(1) The illuminating system,

(2) A lens system to provide magnification and

(3) A means for observing the magnified image.

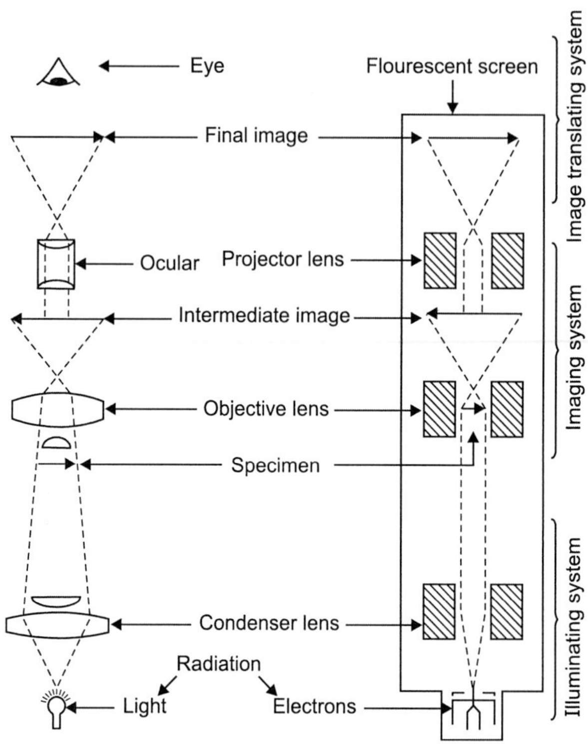

Fig. 2.11: Basic Similarities and Differences in Construction between the Light Microscope and the Electron Microscope

Similarities and Differences in Light and Electron Microscope

Light Microscope	Electron Microscope
1. It uses visible light.	1. It uses electrons.
2. Source of illumination is at the bottom.	2. Source of illumination is at the top.
3. For magnification glass lenses are present.	3. For magnification electromagnetic coils are present. Coils use electricity in order to function as lenses.
4. The lenses are condenser, objective and ocular.	4. It has a condenser and an objective, projector coil.

Light Microscope	Electron Microscope
5. Image can be seen with eye or on photographic film.	5. It can be seen on fluorescent screen or recorded on photographic film.
6. The principles of fixation, staining, embedding, mounting basically similar.	6. Their procedure and requirements differ.
7. Differ in construction, working and procedures.	7. Differ in construction , working and procedures.

Properties of Electrons:

The source of illumination in EM is electron.

1. They are negatively charged sub-atomic particles. When the atoms of metal are excited by sufficient energy in the form of heat, the electron velocity is accelerated to a rate at which electrons leave their orbit, fly off into space and are lost to the atom.

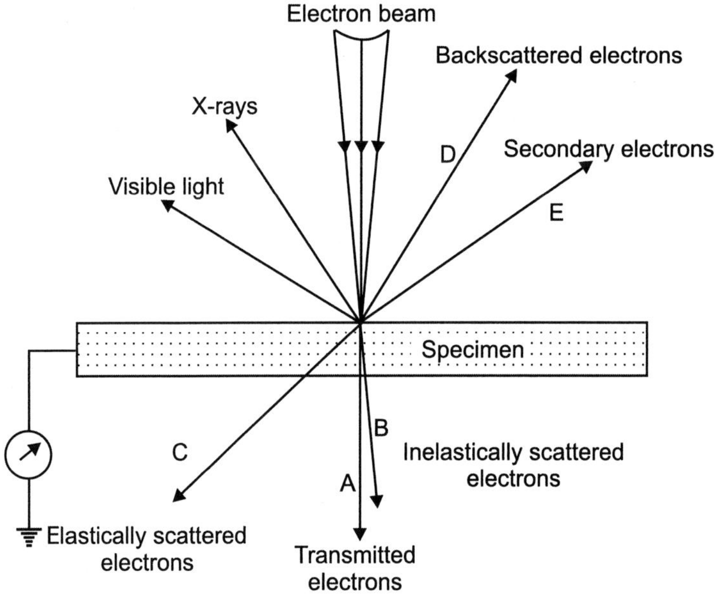

Fig. 2.12: The Behaviour of an Electron Beam. The Transmitted Electrons Forms the Image, the Rest of the Electrons get Scattered

2. The metal tungsten is commonly used as a source of electrons.
3. The atom bounded electrons, when excited by a current of high voltage fly off from the surface of the metal in a continuous steam.
4. The electrons emitted from the electron gun are passed through a collimating (rendering rays parallel) aperture to form a well defined beam.
5. The electrons are readily absorbed and scattered by a different form of matter. A beam of electron, therefore, can be produced and sustained only in high vacuum.
6. Electrons are light wave i.e. they are corpuscular and vibratory in nature and therefore, are used for image formation.

They interact with the atoms of the biological specimens to form the image. The electrons behave in several different ways when interact with the atoms. (Fig. 2.12).

(a) The electrons (A) transmitted from beam pass through specimen on to the fluorescent screen or the photographic film.

(b) Some electrons (B) lose a bit of their energy while passing through the specimen and get deflected from their original axis of the beam. Such electrons are called inelastically scattered electrons.

(c) Some electrons (C) interact with atoms of specimen and get elastically scattered without loss of energy. The electrons deviate more from their axis.

(d) Some electrons (D) get back scattered instead of passing through specimen.

(e) Some electrons get absorbed by the atom of the specimen and instead low energy electrons (E) are emitted. They are called secondary electrons and they are useful for information of image in the scanning EM.

(f) Some electrons dislodge electrons from the atomic shells of atoms in the specimen. Each empty space is filled by an electron from an outer shell. The atoms then emit X-rays of a characteristic wavelength and energy.

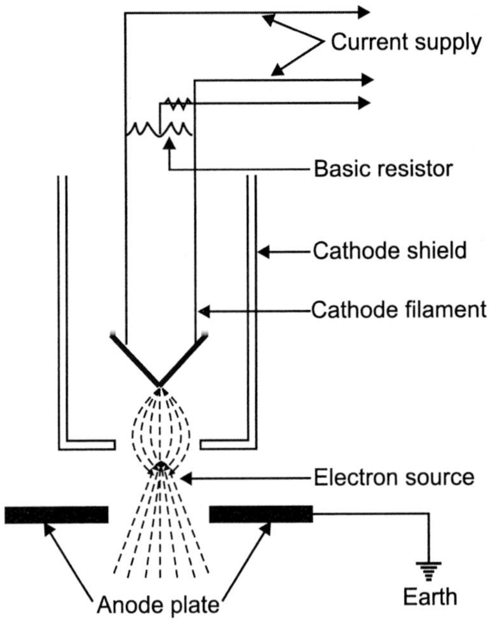

Fig. 2.13: A Self-Biasing Electron Gun

The wavelength of the electrons is based on the voltage of the electric current. It can be calculated by de Broglie's formula.

$$\lambda = \frac{12.3}{\sqrt{V}} \text{ Angstroms}$$

Constructions:

The electron microscope consists of the following parts:

1. **Electron Gun:** It is located at the top of the microscope body and it is the source of electrons. It consists of a hot, tungsten filament. It is source of electrons forming the beam. The tungsten filament is surrounded by negatively biased shield with an aperture through which an electron beam is drawn off to grounded and positive anode below.

2. **Microscope Column:** It consists of an evacuated metal tube. It is fixed at the top of electron gun, a number of electromagnetic lenses, viewing screen and photographic plate. These parts are aligned one above the other. The microscopic column provides a protective shield to the operator from X-rays that are generated when electrons strike the metal surface.

3. Electromagnetic Lenses or Coils: The coil correspond to the condenser, objective and ocular lenses of light microscope. Each coil has coils of electric wire wound on a hollow metal cylinder, designed in such a way that an electric current passing through the magnetic coil produces an axially symmetrical magnetic field in the center of the lens. The magnetic filed forces the electrons to spiral around a central axis. The electron beam passes through the microscope column and deflected by a variable degree depending on the current flowing through the coil of the lens. The microscope is fitted with a variable resistance with a rotary switch control. It controls focusing and image formation in the electron microscope. The magnetic field functions as a magnifying lens. Focusing is also done by adjusting the voltage of the current.

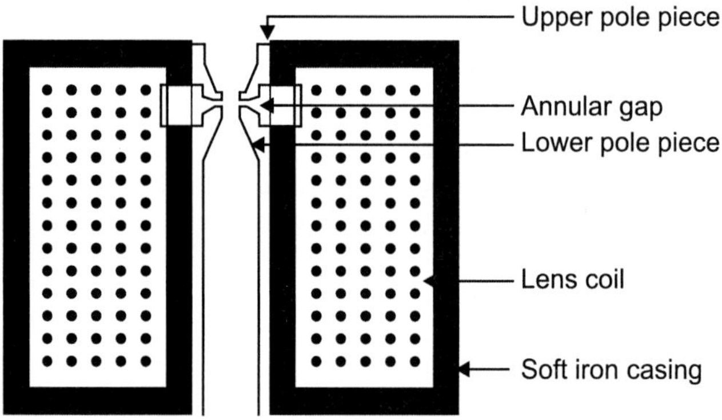

Fig. 2.14: An Electromagnetic Lens with the Pole Pieces

4. Fluorescent Screen: Electrons are harmful to our eyes, the magnified image is observed on a fluorescent screen. The screen is located with a chemical which by its excitation, forms the image as on the television screen.

5. Transformers: The EM requires high voltage current for the electron gun and the electromagnetic coils. For this high voltage transformers which can boost the voltage from 220 V to 50 – 100 kV.

6. Vacuum Pumps: EM requires vacuum inside the microscope column. This is maintained by diffusion pumps or iron pumps.

7. Water Cooling System: EM requires cooling system to prevent overheating of various parts by the pumps and coils. Now-a-days modern electron microscopes are equipped with circulating pump; refrigeration plant and filter system. The size is reduced because of electronic system.

Working: (Fig. 2.15)

The working of EM is very much similar to that of light microscope. The electrons are used for magnification and image formation. Image formation occurs by electron scattering. Electrons strike the atomic nuclei and get dispersed and these dispersed electrons form the image. The electron image is converted into a visible form by projecting on the fluorescent screen.

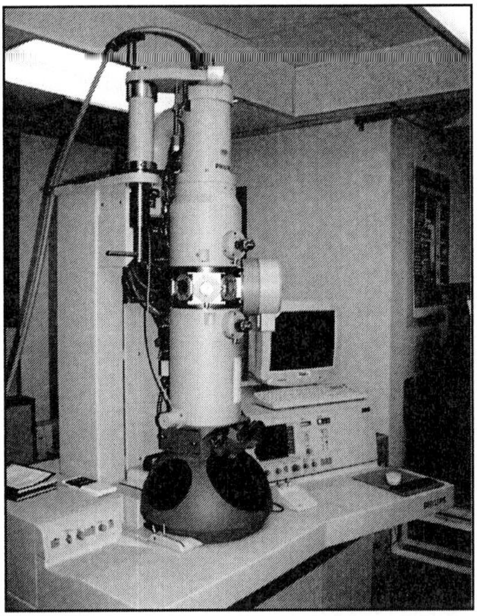

Fig. 2.15: The Electron Microscope

Electrons in the form of beam passes through the condenser coil and fall on the object. They get scattered and transmitted through the object and passes through the objective coil, which magnifies the image of the object. The projector coil further magnifies the image and projects it on the fluorescent screen or photographic film. The image formation occurs when the energy of the electrons is transformed into visible light through excitation of chemical coating of the screen. Those electrons which reach to the fluorescent screen from the bright spots while the areas where the electrons do not reach to screen form dark spots. Electron dispersion is due to atomic nuclei which consists of protons and neutrons. The higher the atomic number the greater is the dispersion. Biological materials are generally with low atomic number. So dispersion of electrons is poor. Thus it results into poor contrast in the image formation. Therefore, to increase the contrast, number of salts with high atomic number are used during the process of fixation or staining.

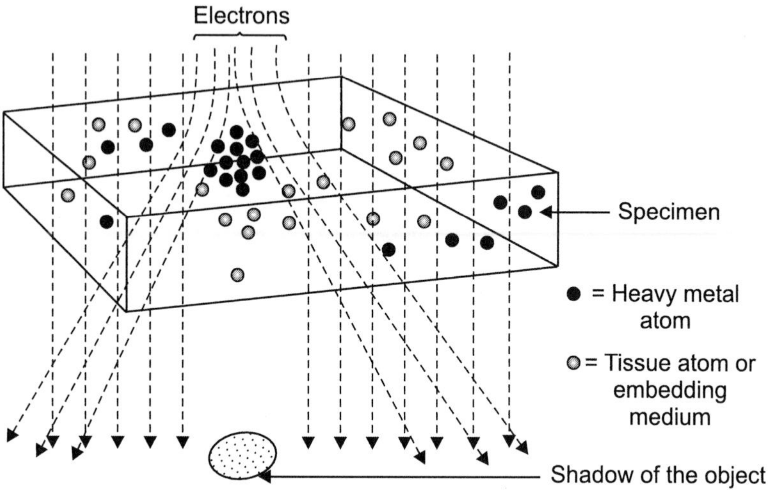

Fig. 2.16: The Principle of Image Formation in Electron Microscope. The Clusters of Atoms of High Density or Atomic Number in the Specimen Cause Scattering of Electrons. As a Result of Scattering, Fewer Electrons reach Corresponding Region on the Screen and Form a 'Shadow' of the Cluster. The Sites where the Electrons reach Unscattered form the Bright Spots

Magnification: The objective and projector coils help in magnifying the image formed in the electron microscope, for getting maximum magnification an intermediate coil is fitted between the objective and projector coils. This coil further increases the magnification of the object obtained with two coils. If the magnification of the objective coil is 100 and projector coils is 200, we get a net magnification of 20,000. Electron microscopes can achieve a magnification of as high a 1,60,000. This magnification can be further increased during microphotography upto 1,000,000 times or more without loss of sharpness.

The resolving power of a microscope is limited by the wavelength of illumination formation of the image. The shorter the wavelength the smaller the detail that can be resolved. The electrons, therefore, can be used with much shorter wavelength for finer details.

For electron microscope sections are cut with diamond or glass knives because ordinary knives are unsuitable for getting thin section. Section of

10 - 50 nm thickness (ultra thin) are used and they are cut with the help of ultramicroscope. Electron microscope in which electrons are allowed to be transmitted through the objects are known as Transmission Electron Microscope (TEM).

High Voltage Electron Microscope:

Now-a-days, it is possible to study living cells with the help of high voltage electron microscopes. In traditional EM, the electricity is used in the range of 50 to 100 kV. But at this voltage the beam of electrons cannot pass through thick sections or living cells. Recently, High Voltage Electrons Microscopes (HVEM) have been introduced which use electricity at a voltage ranging from 500 to 300 kV. Thus, this type of microscope is useful for study of living cells and several microns thick section.

Scanning Electron Microscope:

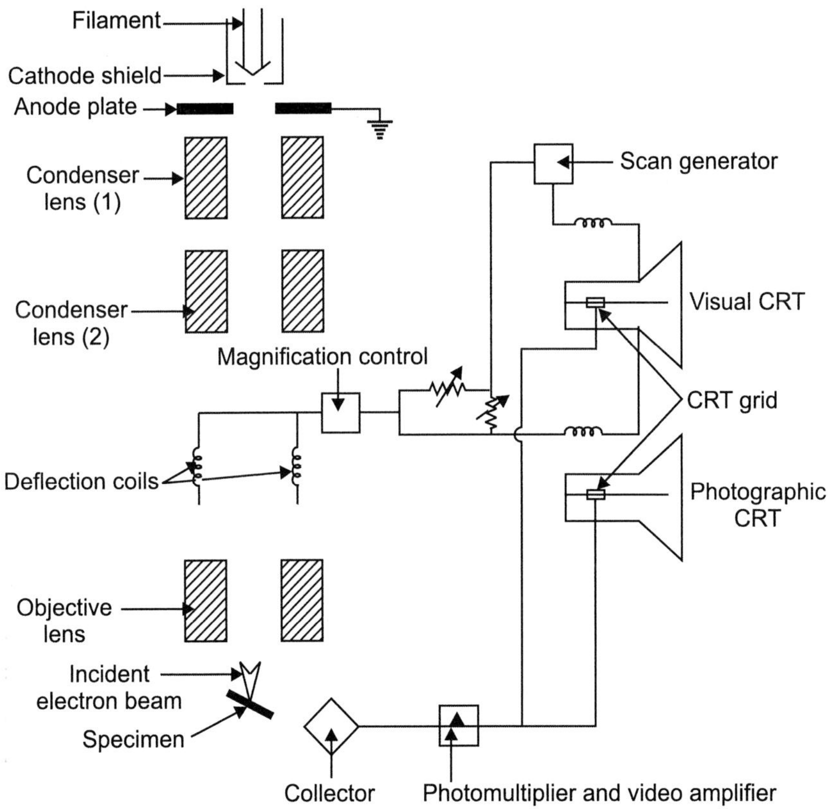

Fig. 2.17: A Diagrammatic Section of Scanning Electron Microscope

This type of microscope is extremely useful for studying surface structure of thick specimens. This method of examining is also called Surface Scanning Electron and Microscopy (SSEM). In this type, the electrons do not form the image by being transmitted, but by getting emitted from the surface of the specimen. The microscope gives a typical three dimensional appearance. The illuminating system is similar to that of transmission electron microscope. The electron beam is, however, compressed with the help of one or more condenser coils, which result in the formation of an electron probe that is nothing but a narrow pencil of electrons. The electrons that fall on the surface of material are called primary electrons. The electron probe is focused on the surface of the specimen. The electrons are emitted as secondary electrons from the surface. The electrons do not pass through the specimen. There is no image formation system in this microscope. The image is formed on the cathode ray tube of the reflected electrons by the television electronics. The three dimensional image of SEM is formed by secondary electrons, but they have no capacity to excite fluorescent screen. In scanning electron microscope the secondary electrons are first collected, amplified and then used for the formation of an image on the phosphor screen of a cathode ray tube.

There are two such tubes one for visual observation and other for photograph. The material to be studied under this microscope is coated with an alloy of gold - palladium and then it is transferred to a grid for support. This helps in producing large number of secondary electrons for obtaining a good picture. The scanning electron microscope has a wide scope in biology for the study of small specimens, surface scanning of the cells, tissues and membranes.

In addition to above, scanning transmission electron microscope is present which is nothing but a modified transmission electron microscope. It is provided with a suitable holder for thin specimens for scanning and an electron detector below the specimen to receive the transmitted electrons to form the image, the image formed in transmission electron microscope is due to the unscattered and inelastically scattered electron which reach the fluorescent screen.

The stereo electron microscopy uses the scanning electron microscope to obtain a three dimensional or stereoscope image of the specimen.

Resolving Power and Numerical Aperture:

Resolving power is defined as the capacity of a microscope or any other instrument to distinguish images of two pointed objects lying very close together. The microscope magnifies as well as resolves the objects seen through it. But resolving power is more important property of the microscope than the property of magnification. If the resolving power of a microscope is low, the image of two closely placed points will overlap only blurred single point will be visible. But when the microscope has high resolving power, these two points will appear sharply distinct.

If two points are more than 0.1 mm apart, our unaided eyes will be able to distinguish them as two distinct objects. If two points are at a distance of more than 0.2 μm, they will appear as two points in the microscope. Thus, the microscope can resolve the objects. However, there is a limit of resolution. (Fig. 2.18)

The resolving power of microscope depends on two factors.

(i) The wavelength of light employed and

(ii) The numerical aperture is normally used to calculate the resolving power of the microscope.

$$R = \frac{0.61\,\lambda}{N.A.}$$

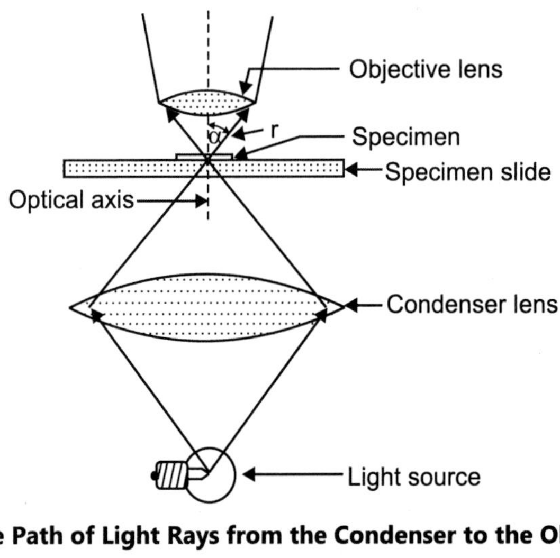

Fig. 2.18: The Path of Light Rays from the Condenser to the Objective Lens. The Half – Angle (α) is between the Optical Axis and the Widest Accepted Ray (r).

0.61 is the constant representing the minimum detectable difference in contrast λ is the wavelength of illumination. Numerical Aperture or N.A. is the light gathering capacity of the objective. N.A. depends on the refractive index of the lens and sin of the semi angle of the aperture as follows:

$$\text{Numerical aperture (N.A.)} = n \sin \alpha$$

where, n is the refractive index of air or liquid between the specimen and lens, sin α is the ≥ sin of the semi angle of the Aperture and α is the half angle of the cone of light entering the objective lens from the specimen. It is important to note that the limit of resolution is inversely proportional to the resolving power. In other words, the higher the resolving power the smaller is the value of the limit of resolution.

Sin α cannot exceed the value of 1 and the refractive index of the optical instrument is constant i.e. it is 1.6 for a 40 × objective and 1.4 for the oil immersion lens (100 ×).

The limit of resolution for the microscope can therefore be calculate as follows:

$$\text{Limit of resolution} = \frac{0.61 \times \text{Wavelength } (\lambda)}{1 \times 1.4}$$

The following table gives relation between N.A. and limit of useful magnification:

N.A.	Approx. limit of useful Magnification
0.1	100 ×
0.25	250 ×
0.50	500 ×
0.66	660 ×
0.00	1000 ×
1.3	1300 ×

When oil immersion is not used the lenses will have N.A. maximum of 1.0. Oil immersion lenses have a maximum N.A. of 1.5.

Normally, the N.A. of low power objective is 0.25 and that of high power objective is 0.66. The value of N.A. is engraved on the objective tube along with magnification value. Thus, N.A value of the objective lenses is closely linked with resolving power of microscope.

2.2 STAINS AND DYES

Staining refers to the use of dyes or colours to render various tissue constituents visible and distinct from one another. Different parts of cells or tissues react differently to colours due to their chemical or/and physical differences. Thus, they become easily visible. The preparation and uses of stains is a separate science in itself.

The constituent parts of cells and intracellular materials are usually transparent and colourless and therefore cannot be differentiated from one another unless there is appreciable difference in their refractive index. It was **Leeuwenhoek (1774 – 1719)** who first used a colouring agent in biological microtechnique, to overcome this difficulty. He was studying mammalian muscle under a single lens microscope. He found his section too transparent for observation and he therefore, soaked it in crocus (saffron) macerated in wine. Iodine was commonly used for the same purpose.

Staining is an auxiliary technique used in microscopy to increase contrast in microscopic specimens. Stains and dyes are used to highlight structures in biological tissues for viewing; with the help of different microscopes. Stains are also used to examine bulk tissues (muscle fibres or connective tissue), cell populations or organelles within individual cells.

2.2.1 Definition

Cell, tissues and microscopic organisms are generally studied under a microscope only after proper fixation and staining. Staining is a method of dyeing the various cell organelles or its components by using certain chemicals. The chemicals used for colouring are called dyes or stains.

2.2.2 Principle of Staining

All the dyes or stains are generally aromatic substances. Stains are classified in two groups – synthetic and natural.

Each stain has two kinds of active chemical groups – chromophoric and auxochromic.

a) Chromophoric Group: This group imparts colour to the stain. Some of the chromophoric groups are nitro group (– NO_2), carboxyl group (– COOH), azo group (– N = N –), indamin group (– N =) and quinoid group (O = 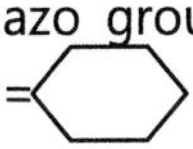 = O).

b) Auxochromic Group: This group dissolves and dissociates in water and helps in the attachment to the cell component. Thus with the attachment of the auxochrome group, the chromophoric group also gets attached giving the cell component the colour of the stain. An example of auxochrome group is hydroxyl group (–OH).

The term staining involves the concept that the dye / stain must become attached to the structure to be stained and remain attached through the various manipulations inherent in the process of mounting. The common reason for the permanent attachment of the dye is the charge on the dye is balanced by the charge on the specimen which is stained. For example, the basic dyes like methylene blue or hematoxylin become attached to those molecules which retain their negative charge (cations) after they have been fixed or sectioned e.g. nucleic acids. So the dyes mentioned above are known as nuclear stains.

Most of the proteins present in the cytoplasm retain positive charge (anion), so that acidic dyes like eosin remain attached to them and therefore these are called cytoplasmic stains.

2.2.3 Classification of Stains

Depending upon the colour, the radical being a cation or an anion or both, stains are classified into three types.

I. Basic Stains (Nuclear Stains)

The chromophoric group of the basic stain is a cation. The net charge on the cation of the basic stain is positive. Such stains have a greater combining capacity at alkaline pH. They are generally used for staining the nucleus (nucleic acids). Basic stains impart colour (stain) to acidic substances and are dye-salts of colourless acids. A substance which shows affinity for basic stain is called basophilic and the property is termed as basophilia. The basophilic substances include the nucleic acids and chromosomes. The examples of basic stains are haematoxylin, methylene blue, methyl green, basic fuchsin, crystal violet, acridine orange, Azure – B, Thionine and Safranin.

Two examples of nuclear stain with reference to composition and importance is given below.

1. Haematoxylin

Haematoxylin is a nuclear stain always used to study the basic differentiation between nuclei and cytoplasm. It is commonly used for both plant and animal tissues. This stain is obtained in a powder form from *Haematoxylon campechianum Linn,* a small leguminous plant that contains haematoxylin. For staining purpose, it is converted by oxidation into haematin either slowly due to atomospheric oxygen (ripening) or rapidly by addition of oxidising agents like sodium iodate, sodium perforate, mercuric oxide, $KMnO_4$ and potassium periodate. It means that haematoxylin ($C_{16}H_{14}O_6$) is not itself a dye. It first needs to be oxidised to haematin ($C_{16}H_{14}O_6$) by exposing it to atmospheric oxygen.

In haematoxylin or haematin, some mordent e.g. (alum) is added to enhance the blue, black or violet staining appearance of the nuclei. This stain is universally used due to high contrast, convenience for photography, easy control of dyeing and the insolubility of colour in aqueous or alcoholic media after staining and in synthetic mounting media.

There are three methods for preparation of haematoxylin. The choice of method depends upon availability of time between preparation and use of stain. If longer time lapse is there; then slow ripening method is better. Its requirement is immediate, then quick oxidation is followed.

A. Delafield's Haematoxylin Composition and Preparation

A - Haematoxylin - 1.0 gm
Absolute alcohol - 6.0 ml

B - Ammonium alum - 10.0 gm
Distilled water - 90.0 ml

Combine A and B and allow the mixture to ripen in air. When the solution has darkened, add:

Glycerol - 25.0 ml
Methyl or ethyl alcohol - 25.0 ml

The final mixture may be allowed to ripen further while tests of its staining activity are being made every week or two. When the activity is judged to be optimum, the solution should be stored in tightly stoppered bottles.

B. Ehrlich's Haematoxylin

Haematoxylin - 2.0 gm
Absolute alcohol - 100 ml

Distilled water	-	100 ml
Glycerin	-	100 ml
Glacial acetic acid	-	10 ml
Potassium alum	-	15 – 20 g

1. Dissolve the haematoxylin powder in alcohol and acid and mix thoroughly.
2. Add other ingredients in order given above, each time ensuring that the solution is mixed well.
3. Add some alum, shake well till maximum dissolution is achieved.
4. Allow ripening for some months in a loosely stoppered large flask kept at a place exposed to sun for at least a few hours every day.
5. Add the remaining alum to the ripened solution.

C. Harris's Haematoxylin

Haematoxylin	-	1.0 gm
Absolute alcohol	-	10 ml
Distilled water	-	100 ml
Potassium or Ammonium alum	-	20 gm
Mercuric oxide	-	0.5 gm
Glacial acetic acid	-	8 ml

1. Dissolve haematoxylin in alcohol in a 500 ml flask with wide mouth.
2. Dissolve alum in water and raise the temperature near to boiling.
3. Pour hot alum solution in the flask containing haematoxylin solution. Mix properly.
4. Boil the mixture and add mercuric oxide to the boiling solution. After frothing switch off the heater.
5. Cool rapidly and filter. Add acetic acid. This is an effective nuclear stain.

2. Acetocarmine (1% solution)

Carmine	-	10 gm
45 % glacial acetic acid	-	1 ml
10 % Ferric chloride	-	5 ml

Preparation of Acetocarmine Stain

Carmine is a basic dye that is prepared from the insect *Coccus cacti.* Dissolve 10 g. carmine (Fisher CS79 – 25) in 100 ml of glacial acetic acid, add boileezers and reflux for 24 hrs. Filter into dark bottles and store at 4°C. This solution can be stored for a long time. Staining can be intensified by adding ferric chloride ($FeCl_2 \cdot 6H_2O$); add 5 ml of 10% ferric chloride solution per 100 ml of 1% acetocarmine.

Acetocarmine stain is generally used to stain plant chromosomes.

3. Methylene Blue

Methylene blue is often used for treating aquariums to prevent fungal infections in fish. It is also a widely used biological stain and indicator. Methylene blue preparations are typically 0.5 to 1.0 w/v solutions of methylene blue in an alcohol / water solution made very slightly alkaline by addition of a few drops of dilute NaOH or KOH. Loeffler's methylene blue is 0.5 g methylene blue powder dissolved in a solution of 100 ml distilled water, 30 ml ethanol and 1 ml dilute KOH(1%).

Methylene blue is suitable for staining cell nuclei and bacteria.

4. Aceto-orcein Stain

Composition

Glacial acetic acid	-	45 ml
Orcein powder	-	2 gm
Distilled/ deionised water	-	55 ml
Hydrochloric acid	-	as per requirement

Preparation

1. Mix 45 ml glacial acetic acid with 55 ml of distilled or deionised water.
2. Slowly bring it to a boil.
3. Slowly add 2 grams of orcein to it.
4. Mix as you add. Keep the solution on a magnetic mixer for 30 mins to ensure complete dissolution.
5. Place a glass cover over the beaker. The heat of the acid will condense on the glass and then move back down into the mixer. This serves to help in the mixing.
6. Use an aspirator to pull the mixture down through the filter.

7. Once cooled, the solution is ready for use. Mix a solution of the stain with 1 N HCl. The ratio is 9 parts stain to 1 part acid.

Aceto-orcein is a very good nuclear stain for both plant and animal nuclei.

II. Acidic Stains (Cytoplasmic Stain) **(April 2014)**

The chromophoric group of the acidic stains is an anion. The net charge on the anion of acidic stain is negative. Such stains have a greater combining capacity at acidic pH. Acidic stains are generally used for staining cytoplasmic components like proteins. Acidic stains are used as salts of sodium or potassium and are neutral or slightly alkaline in reaction. These may contain nitro ($-NO_2$) or quinoid (O = ⬡ = O) groups, as chromophoric groups. The auxochrome groups that combine with the proteins of the cytoplasm may be hydroxyl (–OH), carboxyl (–COOH) or sulphonic ($-HSO_4$) groups. The reaction here is said to be basic. A substance which shows affinity for acidic stain is known as acidophilic and the phenomenon is termed as acidophilia. Acidophilic substances are proteins present as components of cytoplasm.

Most of the proteins found in the cytoplasm, retain a positive charge, so that acid dyes like eosin remain attached to them.

It is believed that cytoplasmic staining is primarily due to proteins or to charged groups on the side chains of amino acids constituting the proteins.

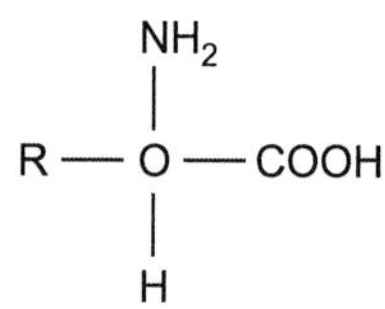

Undissociated amino acid

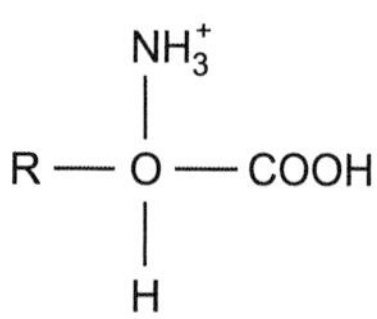

Basic (+)
pH 1.0

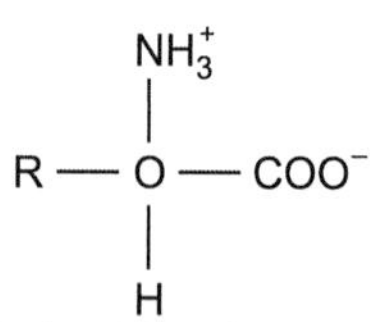

Zwitterion (electrically neutral)
pH 6.0
(Isoelectric point)

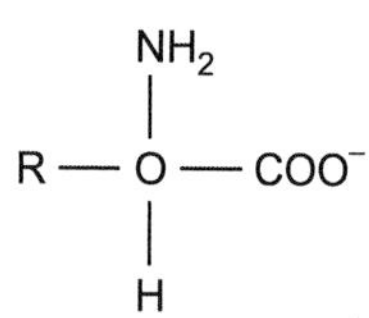

Acidic (–)
pH 14.0

Proteins or polymers (chains) of amino acids contain a terminal (–NH_2) group on one end and a terminal carboxyl (–COOH) group on the other side; in addition, amino acid side chains may have – NH_2 or – COOH groups. They can be either positively or negatively charged, and this charge is pH dependent. As proteins can carry positive or negative charge, they are said to be **amphoteric**.

Isoelectric Point: Isoelectric point is the point where the positive and negative charges are equal. The isoelectric point is approximately pH 6.0; below pH 6.0 the net charge is positive attracting anionic dyes (acidic dyes). Above pH 6.0 the net charge is negative attracting cationic dyes (basic dyes). The substances attracting basic dyes are called basophilic and those attracting acidic dyes are called acidophilic. Isoelectric point is the pH at which the net charge of any protein is neutral.

If tissue sections are placed in a solution below pH – 6.0, the cytoplasmic proteins will develop a predominance of positive charges and will then attract a negatively charged dye such as eosin.

Acid fuchsin, methyl blue, eosin, picric acid, orange G, congo red, bismark brown, fluorescein and aniline blue are the examples of acidic stains.

a) Eosin

It is a xanthene group dye and derived from fluorescein. It is a commonly used stain for staining cytoplasm of the cell. Eosin – Y (Eosin Yellowish) is readily soluble in water and less soluble in alcohol. A 0.2% solution in 95% alcohol is generally used for counter staining the sections in which nuclei are stained blue with haematoxylin. Crystals of thymol or a few drops of formalin are added to prevent the growth of molds. From the stock solution, 1% eosin is usually prepared as a working solution i.e. 1 ml of eosin stock solution in 99 ml water.

Eosin imparts a bright, clear pink colour to the cytoplasmic material, cell membranes and extracellular structures interfering with the blue colour of the chromatin stain.

b) Congo Red

Congo red is the sodium salt of 3, 3'-*(*biphenyl-4, 4'-diyldidiazene-2, 1-diyl*)* bis *(*4-aminonaphthalene-1-sulfonic acid).

It is a secondary diazo dye. Congo red is acidic dye and is water soluble but better soluble in organic solvents such as ethanol. It has a strong non-covalent affinity to cellulose fibres. Congo red was first synthesised in 1883 by Paul Bottiger, who was working for the Friedrich Bayer Company in Elberfeld, Germany.

Due to colour change from blue to red at pH 3.0 – 5.2, congo red can also be used as a pH indicator.

In biochemistry and histology, congo red is used to stain microscopic preparations especially the cytoplasm and erythrocyte. Apple – green birefringence of congo red stained preparates under polarized light is indicative for the presence of amyloid fibrils.

Composition

Congo red	-	0.5 gm
Distilled water	-	95 ml
Alcohol	-	10 ml

c) Acid Fuchsin

Acid fuchsin is used to stain collagen, smooth muscle or mitochondria. It is generally used as a cytoplasmic stain in Mallory's trichrome method. It imparts red colour to collagen fibres and cytoplasm.

Acid fuchsin is generally employed in a 0.2 to 1.0% aqueous solution as the primary stain in Mallory's triple stain. It stains cytoplasm in varying shades of pink.

Composition

Acid Fuchsin	-	3.5 gm
Glacial acetic acid	-	250 ml
Distilled water	-	750 ml

III. VITAL STAINS **(April 2014)**

Vital stains are used for staining living cells. This stain helps in the detection of cell structures and chemical materials in cells. Janus green – B is a vital stain used for staining mitochondria in living cells. This stain is green in colour in its oxidised state but becomes colourless in its reduced state, due to which it is used for staining mitochondria. When living cells are exposed to Janus green – B, the stain gets reduced (colourless) in the cytoplasm but gets oxidised in the mitochondria by the enzyme cytochrome oxidase. Thus, mitochondria appear green in the colourless cells and are maintained in this state as long as cells remain alive.

1) Janus Green – B

This is a basic aniline dye related to both the safranin and azo groups. Solubility of the stain in water is 5.18 % and 1.12 % in alcohol. It can be employed either in strong solution in small amount or a vey dilute solution in normal saline for intravenous injection. It is used for staining nerves intra-vitally and for selectively colouring granules in cells. It is recommended for staining mitochondria in living cells. This stain is also used in various dilutions for fungi and for the flagella of protozoa.

Janus Green – B	-	0.001 to 0.41 gm
Physiological saline	-	100 ml

Chemically, mitochondrial matrix consists mainly of proteins and lipids and the membranes are high in insoluble proteins, phospholipids and insoluble enzyme content. There are many enzymes present in mitochondria. One of these is cytochrome oxidase. Janus green – B when applied in dilute solutions enters the cells and is reduced in the cytoplasm to a leuco-dye. When it enters mitochondria, it is specifically reoxidised to its original colour (green / blue); for this the cells must be kept alive during the experiment.

2) Neutral red

Neutral red is a slightly basic dye and a pH indicator. It is yellow in basic solutions and red in slightly acid solutions. It is primary used in histochemistry as a vital or supravital dye. The dye may be applied to living tissue in dilutions of 1 : 10,000 to 1 : 100,000 in physiological saline solution or it may be applied as a film with or without Janus green. In general, the stain has an affinity for cytoplasmic granules and inclusions in the living cell, staining the nuclei only as the cells die.

Neutral red (or toluene red) stains Nissl substance red. It is usually used as a counterstain in combination with other dye.

2.3 MICROMETRY

1. Study of Units like Micrometer, Nanometer and Angstrom:

Microscopic objects like chromosomes, cells etc. are measured with the help of the compound microscope.

For the accurate measurement of any microscopic object the stage micrometer and ocular micrometer are used in combination. This field of science is called micrometry. The measurement are expressed in micrometers (μm) and are made with the help of an ocular or eyepiece micrometer disc and a stage or object micrometer.

The eye-piece or ocular micrometer is a transparent circular glass disc which is placed in the body of the ocular. The upper surface of the eyepiece micrometer is either engraved with a scale or divided into a number of equal squares by cross – lines ruled upon it. It is placed in between lenses of the eye-piece in such position that the scale can be distinctly on looking through the microscope.

Stage micrometer is a glass slide on which a scale of known intervals is marked. When the scale of the object or stage micrometer is focused; the ocular micrometer is also seen in the superimposed condition. When the two are graduated in the same unit, the ocular micrometer has to be calibrated against the object micrometer. Calibration is done for each objective of the microscope separately, beginning at the lowest magnification and focusing the scale of the state micrometer. Most stage micrometers have a 2 mm long scale divided into 0.01 mm units. The mechanical stage is aligned with the zero reference line of the stage micrometer so that it coincides with the zero line of the ocular micrometer as shown in the Fig. 2.19.

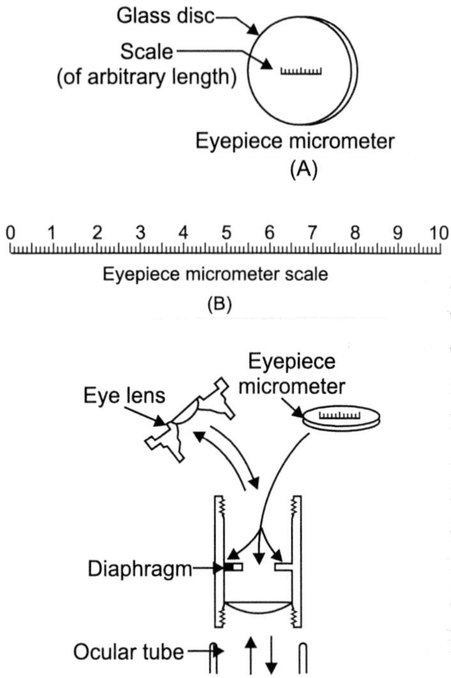

(c) A Longitudinal Section of Eye-Piece to show the Position of Eye-piece Micrometer in it

Fig. 2.19

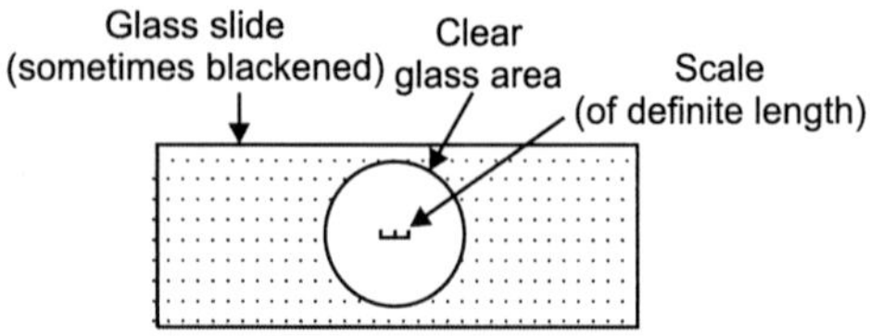

(a) Object Micrometer

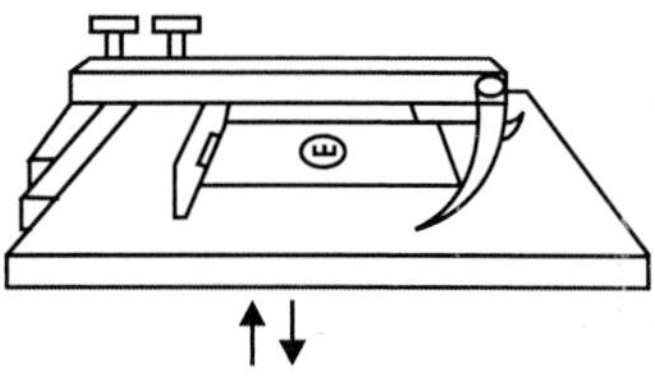

(b) Object Micrometer Fixed on Stage

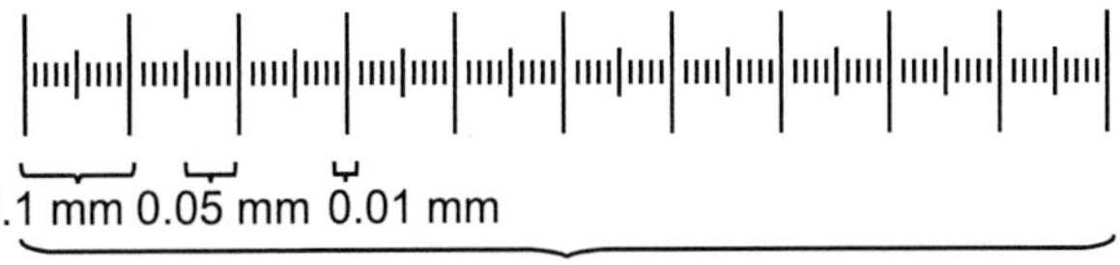

(c) Magnified Object Micrometer Scale

Fig. 2.20

The number of millimeters that correspond to a full scale on the ocular micrometer is first recorded. The distance is then divided by the number of divisions in the ocular micrometer scale and converted to micrometers. In the microscopic field if the full scale on the ocular micrometer coincides with 0.685 mm on the stage micrometer and as there are 100 divisions on the ocular micrometer scale, each division equal 0.685 mm divided by 100 or 6.85 μm perocular micrometer division. This calibration is repeated for each objective magnifcation of the microscope is noted. Whenever, the size of an object is to be a measured the size of the object is set against the ocular micrometer scale in the eyepiece. The reading on the ocular micrometer is noted and converted into the actual size of the object in micrometers (μm).

Unit	Symbol	Value
1 centimeter	cm	1/10 mm or / 100 metre = 10^{-2} metre
1 millimeter	mm	1000 μ or 1/10 cm = 10^{-3} metre
1 micro or micrometer	μ or μm	1000 mμ or 1/1000 mm = 10^{-6} metre
1 millimicron Or 1 nanometer	nm	10A° or 1/1000 μ = 1×10^{-9} metre
1 Angstrom	A°	0.1 mμ or 0.1 mm or 1×10^{-10} metre

2.3.1 Camera Lucida

The camera lucida is an accessory instrument of the compound microscope and is attached to the top of the monocular body tube of the microscope. It is used to draw clear simple and exactly proportionate outline sketches of the objects under study. In the biological work sketches are important and hence this device is commonly used, which was invented in 1807 by W.H. Wollaston.

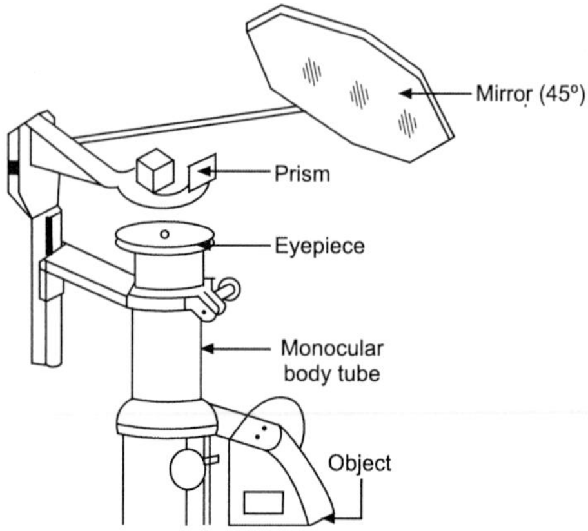

Fig. 2.21: Camera Lucida, Mirror Type

The camera luncida consists of a mirror and reflecting prism which is mounted over the ocular (eye - piece). The surface of the prism is coated with silver plating except in a particular area in the center which coincides in position with the monocular eye - piece. It has an adjustable mirror inclined at 45° both to the desk top and the prism. Light from the paper placed on desk top is reflected by the mirror on to the reflecting face of the prism, which reflects it into the eye of the observer. The light rays passing through the ocular are deflected at 90° by the prism. The rays are further deflected at 90° by the mirror. This arrangement projects an imaginary image on the desk top beside the microscope. If a piece of paper is kept there, the observer can see the superimposed pictures of the paper and the image of the object through the ocular. Observer can see the pencil point. This helps to trace the outline of the image falling on the paper easily by the observer while looking through the eye - piece.

Sketching with the help of camera lucida is not so easy. Certain precautions should be taken, such as avoid bright light passing through object. The bright light makes the pencil point invisible. Even if the paper is bright the object disappears altogether from view. To overcome this problem, many camera lucida are provided with devices like neutral glass filters which can cut the light from either the object or the paper or both. It is advised to use variable resistance in series to the microscope lamp. This helps to reduce the intensity of light. While drawing the figure it is better to used fairly hard pencil which can give precision of line. For record purpose a free hand sketch is sufficient, but for publication in research or scientific journal very accurate and in proportion sketches are required.

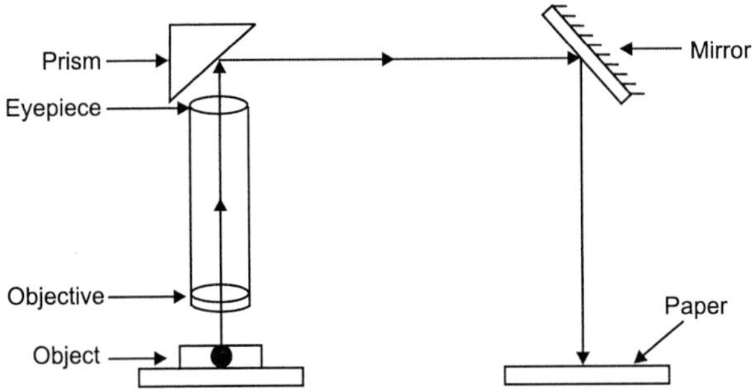

Fig. 2.22: Principle of Camera Lucida

Hence, magnification of the drawing should also be known. For this purpose use of micrometer eye piece to measure the object in various directions is essential. Part of the millimeter scale being drawn with the same set up is used for the drawing. While doing this the angle of inclination of the mirror be should not be altered. If the angle of the mirror is changed the magnification of micrometer scale will change and it will not match with the drawings.

POINTS TO REMEMBER

- The study of different microscopes is called as microscopy.
- The simple microscope is commonly used in laboratory which is nothing but a magnifying glass.
- Improper use of the microscope leads to the loss of clarity of the image.
- Components of commonly used light monocular microscope are support system, illumination system, the magnification system and the adjustment system.
- The electron microscope is an optical instrument which utilizes electrons as a source of illumination for observing very minute object as a great magnification.
- Staining refer to the use of dyes or colors to render various tissue constituents visible and distinct from one another.
- Leeuwenhoek (1774 – 1719) who first used a colouring agent in biological microtechnique, to overcome this difficulty.
- Staining is a method of dyeing the various cell organelles or its components by using certain chemicals.
- Each stain has two kinds of active chemical groups – chromophoric and auxochromic.

- Depending upon the colour, the radical being a cation or an anion or both, stains are classified into three types:
  I. Basic Stains (Nuclear Stains)
  II. Acidic Stains (Cytoplasmic Stain)
  III. Vital Stains
- For the accurate measurement of any microscopic object the stage micrometer and ocular micrometer are used in combination. This field of science is called micrometry.
- The camera lucida is an accessory instrument of the compound microscope and is attached to the top of the monocular body tube of the microscope.

EXERCISE

1. Write the principle and working of phase contrast microscope.
2. Describe the construction and principle of electron microscope.
3. Explain the stage micrometer and ocular micrometer used in micrometry.
4. Describe the construction and principle of camera lucida.
5. Write short notes on:
   (a) Simple microscope
   (b) Principle of light compound microscope
   (c) Principle of phase contrast microscope
   (d) Electron microscope
   (e) Micrometry
   (f) Camera lucida
   (g) Magnification system of compound microscope
   (h) Resolving power and numerical aperture
   (i) Stage and ocular micrometer
   (j) Camera lucida.

❖❖❖

3

CHAPTER

PLASMA MEMBRANE

CONTENTS

3.1 INTRODUCTION

All cells are covered by a very thin membrane called plasmalemma which is not visible under the light microscope. Under the light microscope, the structure seen is called the cell membrane. This cell membrane is made up of plasmalemma and the surrounding cell cement. The plasma membrane may be protected by other coverings e.g. plasma membrane of the egg is surrounded by the vitelline membrane and the jelly layer, both of which are secreted by the ovary.

3.1.1 Chemical Composition of Plasma Membrane

Plasma membrane is made up of lipoproteins. These are special non-bonded combinations of lipids with proteins. Generally, membranes contain about 60% protein and 40% carbohydrates by dry weight.

A. Proteins

On the basis of their function, proteins of the plasma membrane are classified as:

(1) Structural proteins: These proteins form the back bone of the cell membrane and their average molecular weight is 3×10^4.

(2) Enzymes: The structural enzymes which are a major component of the plasma membrane are catalytic proteins.

(3) Carrier proteins: Carrier proteins are also called permeases as they transport substances across the membrane against concentration gradient. Their molecular weight is similar to that of structural proteins.

According to their position in the plasma membrane the proteins of the plasma membrane are of two types :

(1) Intrinsic or integral proteins: These are transmembrane proteins that pass through the plasma membrane. They form channels for the transport of substances into and outside the cells through the plasma membrane.

(2) Extrinsic or peripheral proteins: Proteins which are attached loosely and temporarily to the outer part of the plasma membrane by means of ionic bonds or calcium bridges are known as extrinsic or peripheral proteins.

Plasma Membrane Proteins

(1) Spectrin: It is the most abundant protein. It is a long fibrous molecule that constitutes about 30% of the membrane protein mass. It is useful for maintaining the biconcave shape of the RBCs.

(2) Ankyrin: Spectrin interacts indirectly with integral membrane proteins via this protein.

(3) Glycophorin: It is present on the outer surface of the red blood cells. Its function is unknown.

(4) Enzymes: More than 30 enzymes have been detected in isolated plasma membranes. Those most commonly found are 5'-nucleotidase, Mg^{2+} ATPase, Na^+ -K^+ ATPase, alkaline phosphatase, adenylyl cyclase, RNAse. These enzymes are important in ion transfer across the plasma membrane.

(5) **Glycoproteins:** In erythrocyte membranes, hexose and hexosamine fucose are bound to proteins. They are present on the outer surface of the membrane. These proteins are attached to oligosaccharides, thus forming glycoproteins.

(6) **Integral (Intrinsic) and Peripheral (Extrinsic) proteins:** Spectrin and glycoproteins are examples of these proteins.

B. Lipids

There are two types of lipids associated with plasma membrane. They are called polar lipids. These lipids contain a hydrophilic head and hydrophobic tail and these are connected by glycerol moiety, a sphingosine derivative or a sterol.

In the plasma membrane phospholipids, glycolipids and sterols are present.

Plasma Membrane Lipids

(1) Phospholipids: Most abundant

(2) Cholesterol: A hydrophobic four-membered fused ring structure

(3) Glycolipids: Oligosaccharides containing lipid molecule

They have water loving /hydrophilic or polar end and a hydrophobic / water hating or non-polar end. They influence fluidity of the membrane.

Plasma membrane carbohydrates: Carbohydrates are present in the plasma membrane in the form of glycolipids and glycoproteins on the external membrane surface. The oligosaccharide chains play important roles in cell to cell recognition processes.

3.2 STRUCTURE OF PLASMA MEMBRANE - FLUID MOSAIC MODEL

The fluid-mosaic model of plasma membrane was proposed by ***Singer* and *Nicolson*** in 1972 and it is widely accepted as compared to other proposed models. The important feature of this model is that the biological membranes are considered to be quasi fluid (half fluid) structures in which the lipids and integral proteins are arranged in a mosaic manner. The reason for the wide acceptance of the model is that, it very well explains the properties of cell membrane. The model assumes that there is a peripheral layer of *phospholipid* molecules in which peripheral and integral *globular proteins* are embedded to varying degrees.

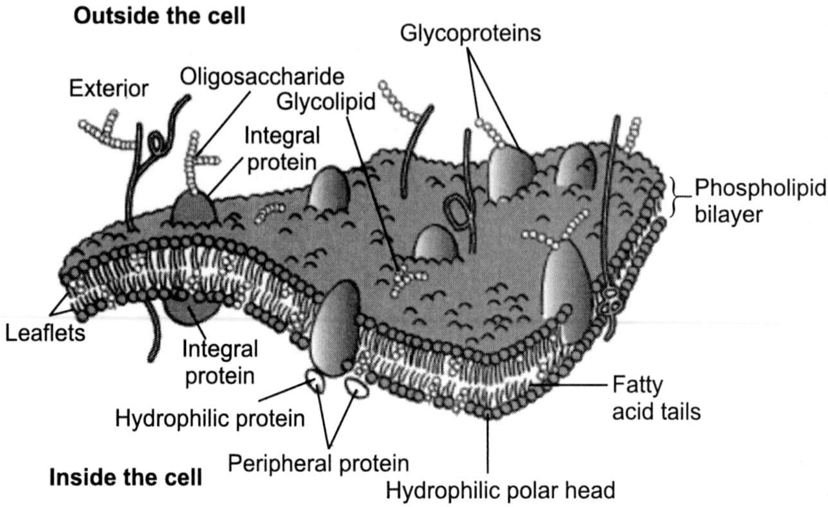

Fig. 3.1: The fluid-mosaic model of Singer and Nicolson

Thus, the proteins are compared to ice bergs floating on a sea of the phospholipid bilayer. During certain immunological experiments, the attachment of antibodies to cell surface exposed integral membrane proteins such as glycophorin which led to the realisation that membrane proteins are not fixed within the lipid bilayer but are free to move laterally, like ice bergs floating or submerged in a sea of lipids. This observation inspired the name *fluid mosaic model,* coined by Singer and Nicolson. This model considers that lipid-protein association to be *hydrophobic* but the *Danielli-Davson* model assumes *hydrophilic* bonding between lipids and proteins. The fluidity of the membrane is the result of this hydrophobic interaction. The phospholipids and many intrinsic proteins are *amphipathic* molecules, i.e. both hydrophilic and hydrophobic groups occur within the same molecule.

The globular proteins which occur in the membrane are of two types, namely, *extrinsic* or *peripheral* proteins and *intrinsic* or *integral* proteins. The peripheral proteins are soluble and readily dissociate from the membrane. They are present entirely outside the lipid bilayer, whereas the integral proteins are relatively insoluble and dissociate with difficulty. Some proteins may partially penetrate through the surface of the lipid bilayer. While others may penetrate directly through the lipid bilayer.

Such proteins are in contact with the aqueous solvent on both sides of the membrane. The integral proteins are amphipathic having both hydrophilic and hydrophobic groups. Their hydrophobic polar heads protrude from the surface of the membrane, while non-polar regions are embedded in the interior of the membrane. The integral proteins are capable of lateral diffusion in the lipid bilayer.

3.3 TRANSPORT ACROSS MEMBRANES

The main function of the plasma membrane is to regulate the flow of material in and out of the cell called transport. This transport of material is regulated by the size of pores present in the plasma membrane. Plasma membrane is semi-permeable which allows the passage of a solvent but not of all solutes. The plasma membrane, which permits one substance to pass through more easily than another is called selectively permeable.

Transport of metabolites through plasma membrane takes place in four ways

(1) Passive Transport : (i) Simple diffusion
(ii) Facilitated diffusion

(2) Active Transport : (i) Primary active transport
(ii) Secondary active transport

3.3.1 Passive Transport

(1) Simple Diffusion: The simplest way for a solute to move from one side of the membrane to the other is simple diffusion. It is the unassisted net movement of a solute from a region where its concentration is higher to a region where its concentration is lower and concentration gradient disappears as diffusion occurs. There is no stereo specificity and the process is slow. ATP is not utilised.

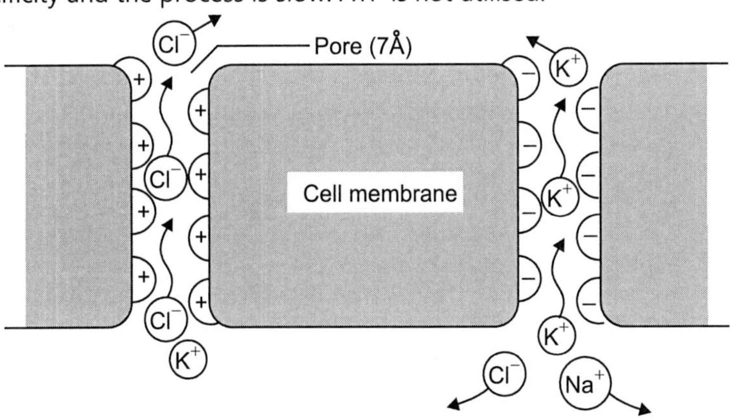

Fig. 3.2: Simple Diffusion

(2) Facilitated Diffusion: In this type of diffusion, energy is not required and it takes place along the concentration gradient. However, it differs from simple diffusion in some respects, (i) It requires carrier for transport of the metabolite across the membrane (ii) It is a stereo specific process i.e. only one isomer is transported.

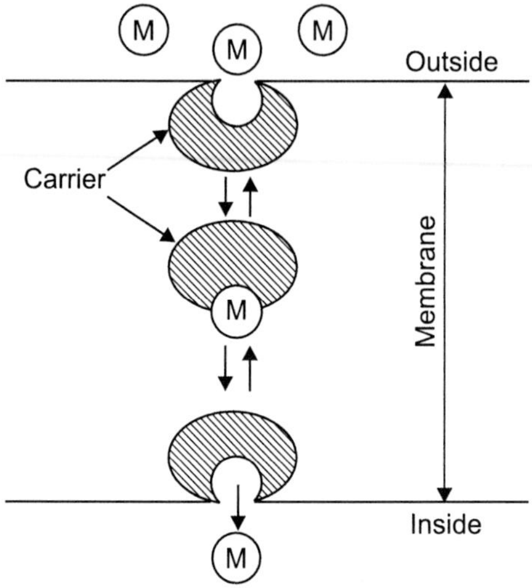

Fig. 3.3: Facilitated Diffusion

The role of carrier protein is simply to facilitate the diffusion of a polar or charged solute. Carriers are proteins with relatively low molecular weight (9 to 40,000) but they are highly selective. The metabolite binds to the carrier protein at the outer surface of the membrane to form carrier metabolite complex. This diffuses along the concentration gradient i.e. from high concentration to low concentration regions.

The metabolite is then released towards the inner surface of the membrane as there is low concentration of metabolites on the inner side. It is continued as long as a concentration gradient is developed. The transport of glucose into the erythrocytes is a good example of facilitated diffusion.

3.3.2 Active Transport

Diffusion of substances takes place across the membrane from high to low concentration. But membranes are also able to affect the transport of solutes up the gradient i.e. in the direction of increasing concentration. Normally, cells have a higher concentration of K^+ ions, which is maintained at all times when the cell is at resting potential. To do so, the cell has to continuously translocate a solute or K^+ ions into the cell by active transport mechanism at the expense of metabolic energy. Most of the inorganic ions, amino acids and carbohydrates are actively transported across the membrane. Cells always translocate essential nutrients from the extracellular environment to the interior of the cell, irrespective of the concentration in the external medium.

This is essential in order to maintain a steady state. In many cells, equilibrium of diffusion can be adjusted by coupling energy yielding reactions, so that the cells are able to translocate solutes or metabolites against the concentration gradient. In addition, it allows various substances, such as secretary products and waste materials, to be removed from the cell or organelle even when their concentration outside is greater than that inside the cell. Third, it enables the cell to maintain constant, non-equilibrium intracellular concentrations of specific inorganic ions, notably K^+, Na^+, Ca^{++} and H^+.

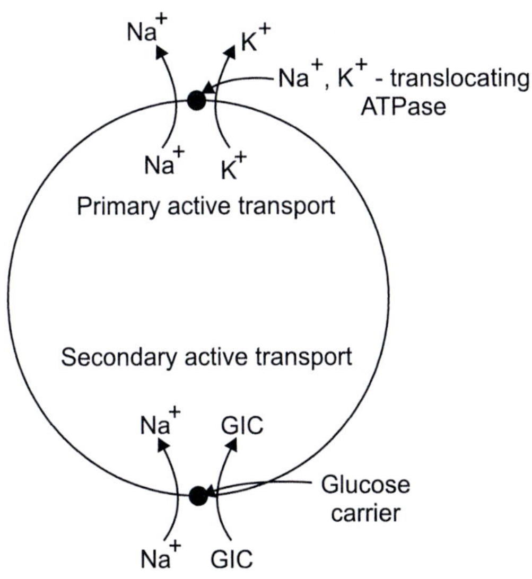

Fig. 3.4: Active Transport

The membrane proteins involved in active transport are often called pumps which selectively transport specific components, molecules or ions from one fluid mass to another. The important features of active transport is that it has directionality. An active transport system that transports a solute across a membrane in one direction will not transport that solute actively in the other direction. Active transport is therefore said to be *unidirectional* or *vectorial.*

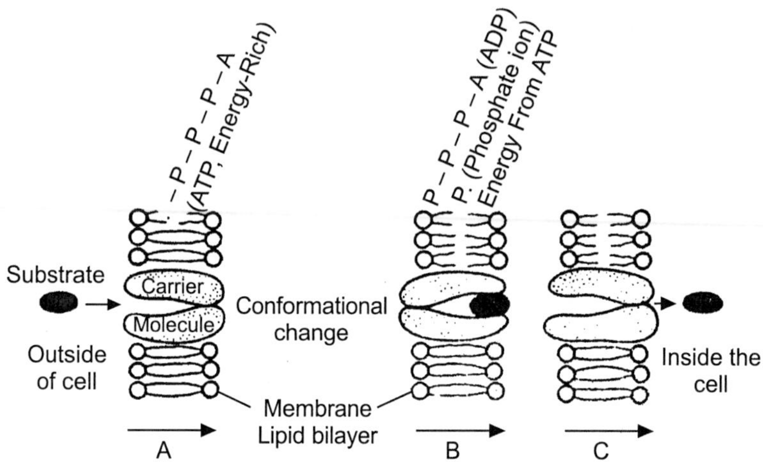

Fig. 3.5: Revolving door model of active transport

Depending on the energy source, active transport is regarded as being either direct or indirect. In direct active transport, the accumulation of solute molecules or ions on one side of the membrane is coupled directly to an exergonic chemical reaction, most commonly the hydrolysis of ATP. Transport proteins driven directly by ATP hydrolysis are called transport ATPases or ATPase pumps.

Indirect active transport on the other hand, depends on the co-transport of the twosolutes, with the movement of one solute down its gradient driving the movement of the other solute up its gradient.

In most cases one of the two solutes is an ion (usually Na^+ or H^+) that moves exergonically down its electrochemical gradient driving the concomitant transport of the second solute such as a monosaccharide or amino acid against its concentration gradient.

Cell membranes allow water and non-polar molecules to penetrate by simple physical diffusion. However, cell membranes are permeable to various polar molecules, such as ions, sugars, amino acids, nucleotides, and many cell metabolites that pass across artificial lipid bilayers very slowly. It is now known that specific membrane proteins are responsible for transferring such solutes across cell membranes. These proteins referred to as membrane transport proteins, occur in many forms and in all the types of biological membranes.

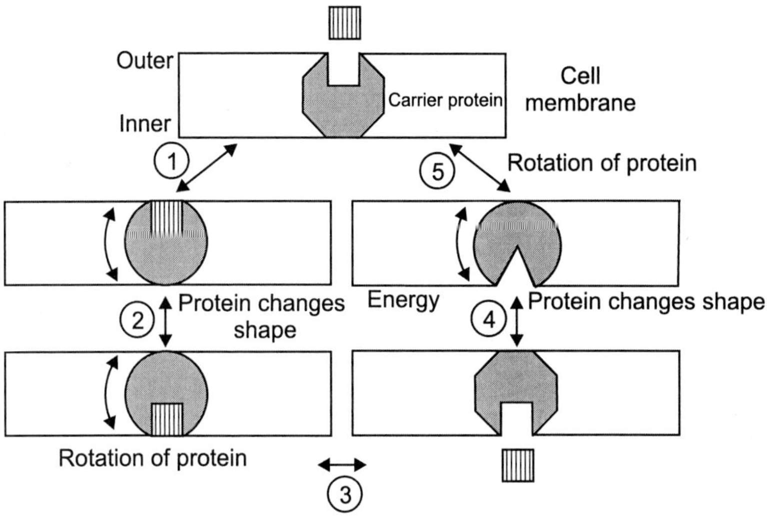

Fig. 3.6: The revolving door model of active transport

Some transport proteins simply transport one solute from one side of the membrane to the other, they are called *uniports*. Others function as co-transport systems, in which the transfer of one solute depends on the simultaneous or sequential transfer of a second solute, either in the same direction is called *symport* or in the opposite direction is called *antiport*.

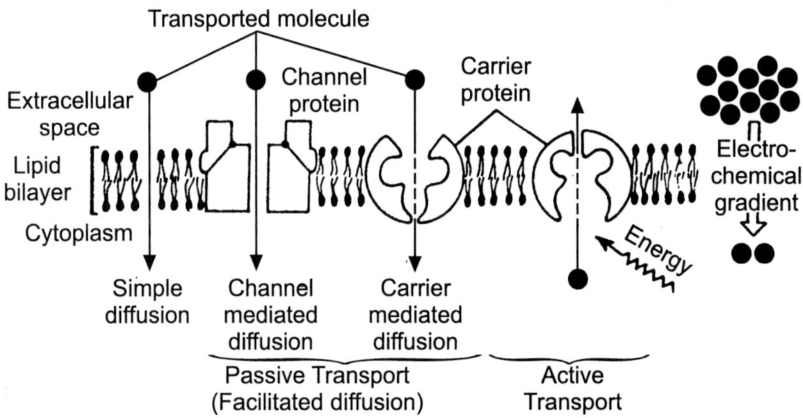

Fig. 3.7: Summary to show difference between different types of diffusion and transport of substances through plasma membrane

3.3.3 Pinocytosis

Intake of fluid material into the cell by the formation of pinocytic vesicles or pinosomes is called pinocytosis (Greek Pinetin-to drink) and has been described first by Lewis (1931). It is a kind of cellular drinking by the cell. This mode of intake of nutrients has been demonstrated in endothelial cells and erythroblasts. Fluids in contact with the plasma membrane move through narrow channels formed temporarily as deep invaginations into the cell interior. From the tips of these channels small vesicles containing the fluid droplets are budded off, which empty their contents in the cytoplasm. The vesicles are called pinosomes. Pinocytosis is found prominently in protozoa and best studied in amoeba. In amoeba, periodic pinocytosis has been observed under light microscope.

The process of pinocytosis is said to be mediated by some kind of inducer mechanism, which triggers the formation of invaginations in the plasma membrane.

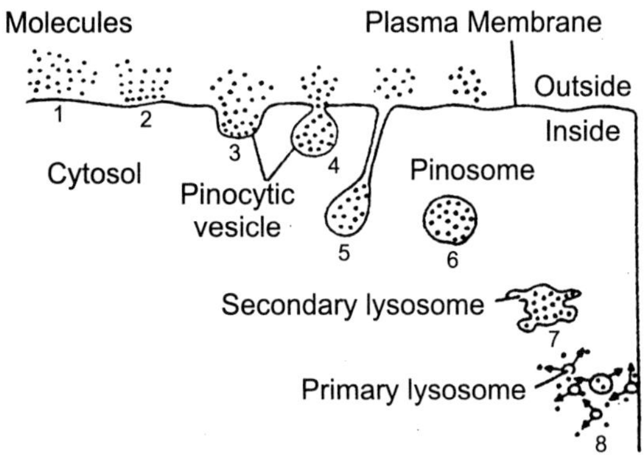

Fig. 3.8: Stages of Pinocytosis

Proteins in solution may act as good inducers as compared to acidic substances like DNA, RNA and fatty acids. All the substances surrounding the cell membrane may not be engulfed, hence the process is highly selective. Surface of the plasma membrane has specific receptor sites, to which only certain materials may be bound and carried through invaginating channels. Endothelial cells lining the capillaries may be cited to prove this point. Materials from the blood are transported into the interstitial space through small vesicles that are formed at the cell border. These vesicles loaded with materials are then transferred to the other side of the cell, where they fuse with the cell membrane and discharge the contents outside.

3.3.4 Phagocytosis

Engulfment of solid particles with the help of plasma membrane has been observed in variety of cells such as leucocytes, endothelial cells, fibrocytes and splenocytes.

The cell ingests or swallows foreign bodies, bacteria, harmful matter and other inert substances and the process is called phagocytosis (Greek Phagein = to eat, kytos = cell).

It is found commonly in protozoans and among certain cells of metazoans. Phagocytosis has been observed in *Amoeba* for obtaining nutrition. The phagocytic cells in tissues of mammals are generally called macrophages, which are either mononucleate or multinucleate and distributed in a variety of mammalian tissues. The particles become absorbed (adheres) at the surface of membrane and later on they are taken into the cytoplasm by infolding of the plasma membrane which are soon pinched off forming vesicles consisting of particles enveloped by a membrane. These vesicles are called intracellular vesicles or phagosomes. The phagosome fuses with a primary lysosome to form a secondary lysosome in which food material is digested by acidic enzymes.

In metazoan, it is a method of defense against foreign bodies like bacteria, dust and various colloids. Ingestion of small colloidal particles is called ultraphagocytosis.

During the process of phagocytosis, a number of events are observed. When there is material near the phagocyte, the cell relays a signal to other cell to make presence known. Once the signal is relayed, other phagocytes pick it up and start moving towards the material to be engulfed. Exact nature is not known, but it is presumed that there is some kind of chemical interaction called chemotaxis between cell and the particulate material. They act on the macrophage surface to mediate a response.

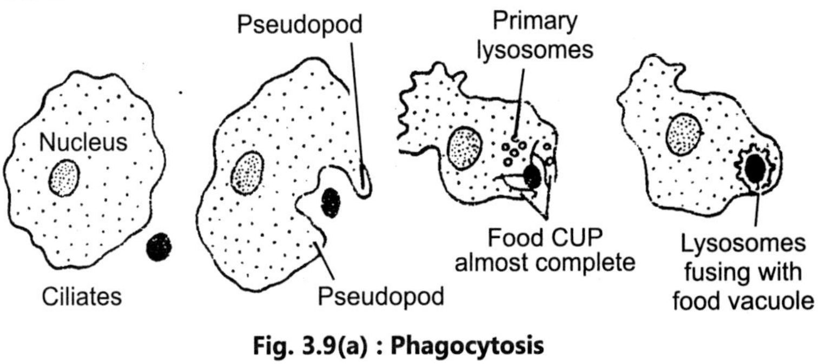

Fig. 3.9(a) : Phagocytosis

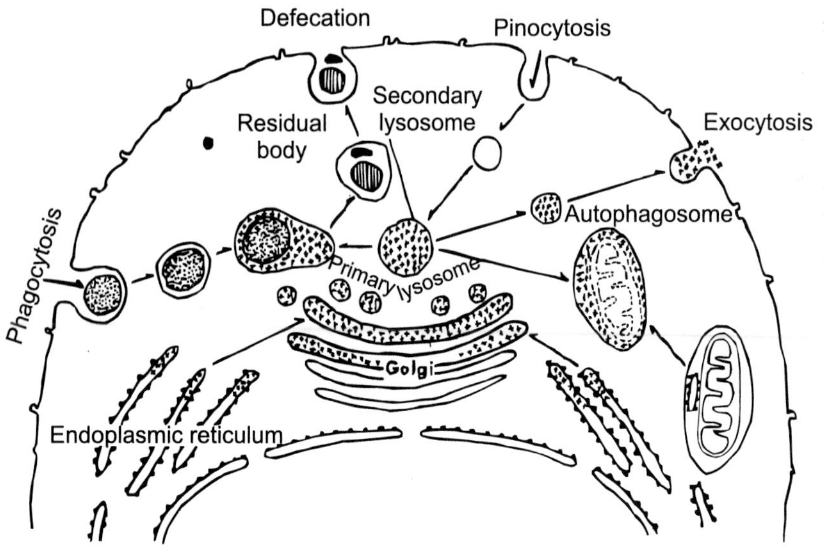

Fig. 3.9 (b): Phagocytosis

3.4 OTHER FUNCTIONS OF CELL MEMBRANE

(1) Transport: The membrane which allows the passage of a solvent but not of a solute is called as a semipermeable membrane. A membrane which permits one substance to pass through more easily than another is said to be selectively permeable. Thus, the membrane allows the passage of both solvent and solute. Molecules diffuse from a region of greater concentration to a region of lower concentration. The difference in the concentration levels is called the concentration gradient. Osmosis is the diffusion of water (or a solvent) across a membrane in response to the concentration gradient. Transport of molecules across the membrane may be active or passive. In active transport, molecules usually move from regions of low concentration to regions of high concentration i.e. against concentration gradient. Thus, it requires energy and depends upon ATP supply. In passive transport, the passage of molecules occurs through the membrane, from a region of high concentration to a region of low concentration. Transport of metabolites across the membrane along the concentration gradient and without the use of a carrier molecule is called simple diffusion.

(2) Cell recognition and adhesion: The mammalian leucocytes recognise foreign cells like bacteria and engulf them by phagocytosis. The sites for cell recognition are known to lie on the surface of the plasma membrane.

(3) Antigen specificity: The glycoproteins on the surface of the cell membranes determine the antigen specificities of the cell. The different blood group systems are all based on the antigenic properties of erythrocytic cell membranes. The ABO blood group system is based on the relationships between antigens on R.B.C. and antibodies in blood serum. The antigenic determinants on the surface of erythrocytes are mainly glycolipids.

(4) Hormone transport: Hormones control cellular metabolism. The cell membrane contains receptors which recognise specific hormones and convey the information to the interior of the cell. This stimulates a change in the metabolism of the cell.

(5) Oxidative phosphorylation: The inner membrane of the mitochondria and the plasma membrane of bacteria contain the electron transport chain which plays an important part in cell respiration. This chain consists of a series of components which can transfer electrons. Electrons from substrates pass down the chain resulting in reduction and oxidation of each component. During this process there is a decrease in free energy, about half of which is utilised for the synthesis of ATP. This process is called oxidative phosphorylation.

(6) Endocytosis and exocytosis: In certain cells, endocytosis is an important activity in which material is transported into the cells by formation of vesicles. It includes two processes namely phagocytosis and pinocytosis. Exocytosis is a process in which membrane lined material is removed from the cell.

(7) Chemoreception: Molecules associated with the cell membrane respond to a variety of stimuli. Chemoreception or the response to chemical stimuli is shown by a variety of organisms from bacteria to mammals.

(8) Transmission: The transmission of nerve impulse takes place at the membrane surface of nerve cells.

3.4.1 Membrane Receptors

Membrane receptors are macro-molecules that have the dual function of recognising a chemical signal and initiating a biological response. Several cellular regulatory agents (e.g. peptide hormones, neurotransmitters) are called ligands. They act as chemical signals, which are recognised by such receptors. At the receptor site there is a specific type of binding that involves a ligand receptor interaction. Several kinds of receptors produce a response by interacting with the enzyme adenylate cyclase within the membrane. As a result, cyclic AMP is produced from ATP. Several receptors that are specific for different ligands may act on a single type of adenylate cyclase. The important point is that the effect of the ligands is not additive, in other words, the system behaves as though several receptors are coupled to one molecule of enzyme. To explain this phenomenon Cuatrecasas and others have postulated *mobile hypothesis* which is based on fluidity of the membrane.

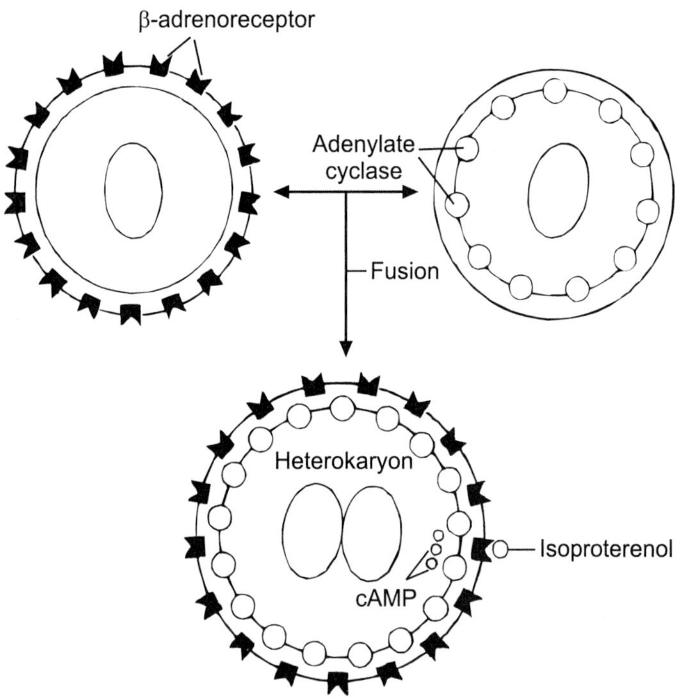

Fig. 3.10: Cell containing the β-adrenergic receptor fused with another carrying only adenylate cyclase. The heterokaryon contains both components and can now couple under the action of the specific ligand isoproterenol resulting in the production of cAMP.

According to this hypothesis, the lateral diffusion of receptors and enzyme within the plane of the lipid bilayer allows several receptors to couple with a single adenylate cyclase.

This hypothesis is confirmed experimentally by fusion studies. Cells having the β–adrenergic receptors in their membrane were fused with others containing only adenylate cyclase. The resulting hybrid cells contained both macromolecules, enabling coupling between the receptor and adenylate cyclase to occur. In these hybrid cells, under the influence of the ligand (in case of isoproterenol) there was production of cAMP, the original hybridised cells were unresponsive to the ligand.

3.4.2 Microvilli

The plasma membrane shows a number of specialisations or modifications on cell surfaces. The narrow invaginations or finger like projections of plasma membrane at the cell surface are known as *microvilli*. The columnar epithelial cells of the intestine are involved in active absorption of digested food. The microvilli lack the internal filaments or tubules found in cilia. The microvilli of intestine are 0.6 to 0.8 μm long and only 0.1 μm in diameter. Microvilli increase the effective surface area for absorption. A single cell may contain as many as 3000 microvilli. In case of intestinal epithelial cells, a dense mat of delicate, branching, tangled filaments cover (fuzzy coat) the surface of microvilli which appear to resist attack from digestive enzymes. It is composed of glycoprotein macro-molecules.

The core of microvilli is traversed by longitudinal microfilaments. These filaments contain actin (a protein) that is attached at the tip of microvilli by a α-actin. They are useful for contraction of microvilli. There are other filaments of protein called *fimbrin* and *villin*. These bind and crosslink F-actin molecules.

At the base, they are joined together by a transverse network of microfilaments extending between the junctional complexes forming a terminal web or bar. In terminal web, besides actin, it also contains myosin, α-actinin, tropomyosin and spectrin.

In the poison gland of scorpion, modified microvilli are found, where soon after origin, the microvilli branch into a large number of secondary branches called dendric microvilli.

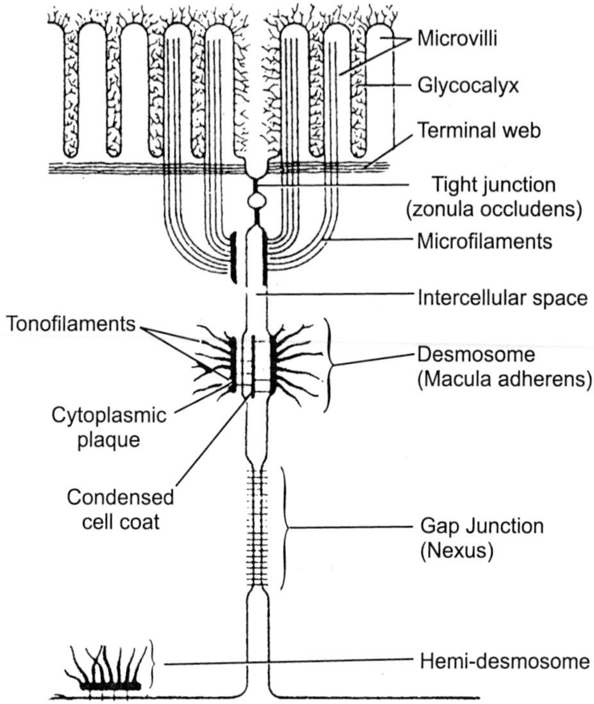

Fig. 3.11: (a) Intestinal wall showing microvilli, tight junction, desmosome and terminal web

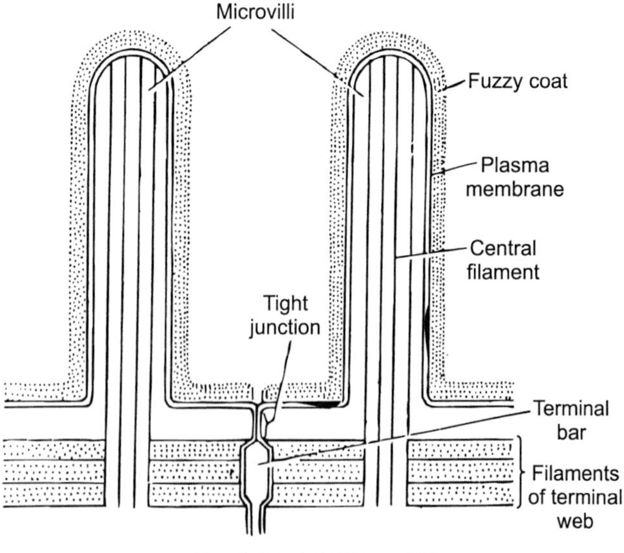

Fig. 3.11: (b) Microvilli

Caveolae: These are cave like invaginations of the cell surface, first named by **Yamada** in 1955, which form blind tubular pouches, opening to the extracellular spaces. These may be small as found in vertebrate capillary endothelial cell, but more definite functional attributes can be assigned to many caveolae. Secretion and uptake of materials into the cell by endocytosis, many small caveolae and vesicles may be secretory in nature.

3.5 CELL JUNCTIONS

Sometimes characteristic junctional complexes have been shown operating for keeping the adjacent cells together. Electron microscopic studies have shown that these junctional complexes are broadly identified as

(a) Tight junctions (*Zona Occludens*)

(b) Gap junctions (*Nexus*)

(c) Desmosomes which are

(i) Belt desmosomes or intermediary junctions (*Zona adherens*)

(ii) Spot desmosomes comprising of *Macula adherens*

(iii) Hemidesmosomes

(iv) Septate desmosomes

3.5.1 Tight Junctions (Zona Occludens)

It is located at the distal region between the adjacent cells where actual fusion of the outer membrane of the adjoining cells occurs. As a result, a single intermediate layer is formed that produces an apparent five layered unit at that region. It acts as seal with no intracellular space. Tight junctions are abundant in the collector tube of kidney. It appears as a network of ridges on the cytoplasmic half of the membrane, with complimentary grooves in the outer half. The ridges appear to be composed of two rows of particles, each one belonging to the adjacent cells.

The lines of these particles produce the sealing and therefore have been named as sealing strands. Tight junctions act as barriers to the diffusion of macromolecules or lipids in the bilayer.

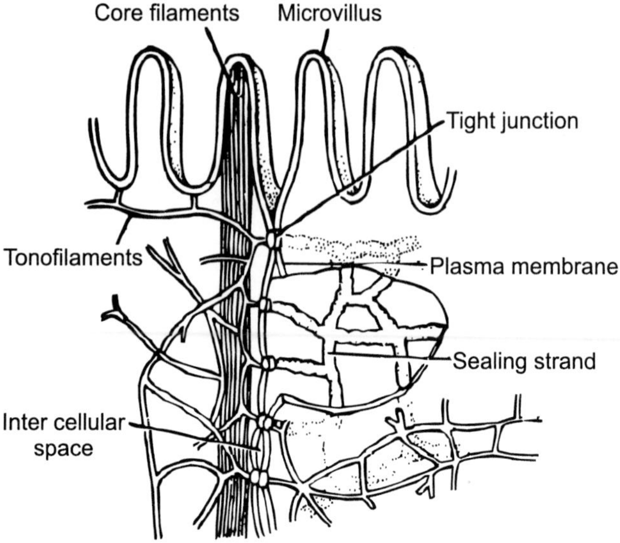

Fig. 3.12 (a): Diagram of spot desmosome

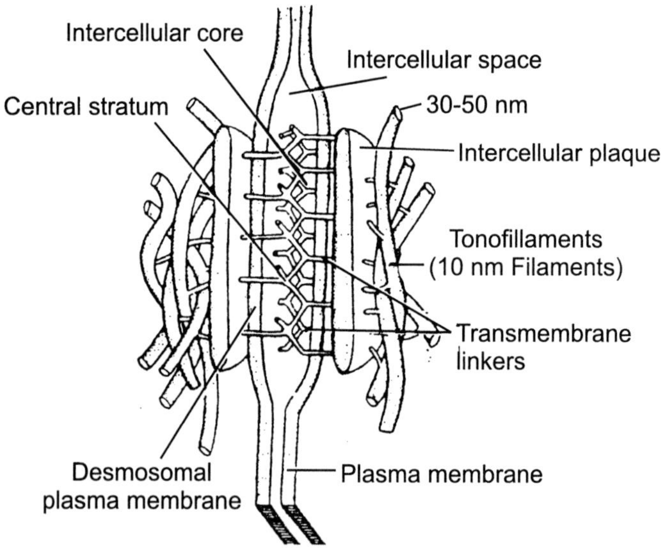

Fig. 3.12 (b): Detailed structure of a tight junction

3.5.2 Gap Junctions (Nexus of Fascia Occludens)

These are called as communicating junctions located at specialised region of contact between opposed plasma membranes of the adjacent cells. It permits direct transfer of ions and other small molecules between cells without traversing the extra cellular space. The gap junction is characterised by a 2-4 nm (20 Å) space between the two membranes. The gap is filled with densely packed protein particles each of which appears to be a channel that connects the two cells. In tangential section, gap junctions show

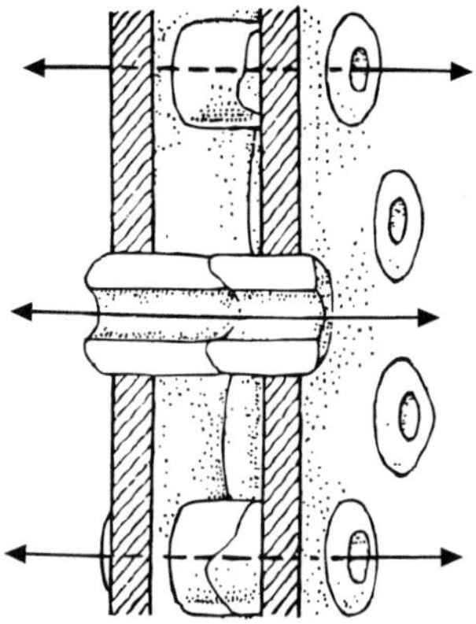

Fig. 3.13: Gap Junction

hexagonal array of 8-9 nm particles. At the centre of each particle there is a channel. They represent regions in which there are junctional channels (1.5 to 2 nm in diameter) through which ions and small molecules can pass from one cell to other. Cells having gap junctions are electrically coupled i.e. there is a free flow of electric current carried by ions.

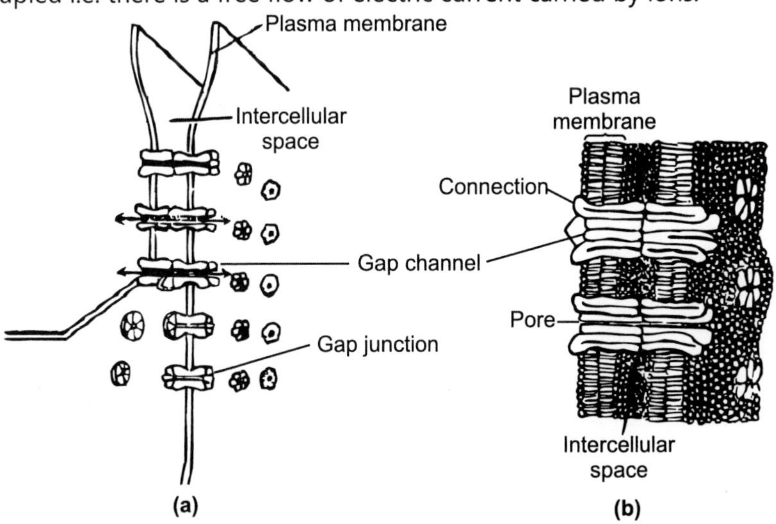

Fig. 3.14: (a) Gap junction between two cells (b) Finner structure of the connexon or junctional unit

The macromolecular unit structure of gap junction has been named the 'connexon' which appears as an annulus of 6-subunits surrounding the channel. It is believed that the sliding of the subunits causes the channel to open and close. The permeability of the channel is regulated by Ca^{++} and ATP provides the energy. Cyclic AMP plays a regulatory role in the formation of gap junction.

3.5.3 Desmosomes

Desmosomes are button like points of strong adhesion between adjacent cells in a body tissue. This cell-cell adhesion gives the tissue structural integrity, enabling the cells to function as a unit and to resist stress. Desmosomes are found in many tissues but they are especially abundant in cells of skin, heart muscle, and the neck of the uterus. Desmosomes form early in embryonic development and play an important role in maintaining cell position during development.

The plasma membranes of the two adjacent cells are aligned in parallel, separated by a space of about 25-35 nm. The extracellular space between the two membranes is called the desmosome core. A thick plaque is found just beneath the plasma membrane of each of the two adjoining cells. The desmosome core is filled with filaments and granules of proteins called desmocollins and desmogleins. These are cadherins that interact with the plaque at the inner membrane surface and mediate cell-cell adhesion at the outer membrane surface.

Cadherins :

The second category of cell-surface glycoproteins are collectively called **cadherins.** Cadherins are of three subtypes such as:

(i) E-cadherin

(ii) N-cadherin and

(iii) P-cadherin

The first category of cell surface glycoprotein is known as Neural Cell Adhesion Molecule or N-CAM which operates in the early embryonic development and is found on the surface of nerve cells and glial cells and causes them to stick together by C^{++} independent mechanism.

E-cadherin is found on many types of epithelial cells, N-cadherin on nerve, heart and lens cells and P-cadherin on cells in the placenta and epidermis. These three cadherins are homologous transmembrane glycoprotein and are made of 700 amino acid residues. They play important roles in later stages of vertebrate development because their appearance and disappearance correlate with major morphogenetic events in which tissues segregate from one another.

Unlike E-cadherin, desmocollins and desmogleins interact heterophilically across the intercellular space. Like other cadherins, linker proteins bind to their cytosolic tail and link them to the cytoskeleton. The β-catenin family protein *plakoglobin* binds to the desmocollin, plakoglobin in turn binds to a plakin protein called *desmoplakin*. Desmoplakin further attaches to *tonofilaments* which are composed of intermediate filaments such as *vimentin, desmin* or *keratin*.

Tonofilaments extend inward from the plaque, anchoring the desmosome in the cytoplasm. The main protein present in the tonofilaments is the intermediate filament keratin, desmin, or vimentin, depending on the cell type. Each desmosome has anchoring sites for intermediate filaments in the cells on both sides of the junction, thereby linking the intermediate filaments of adjacent cells and forming continuous cytoskeletal network throughout the tissues.

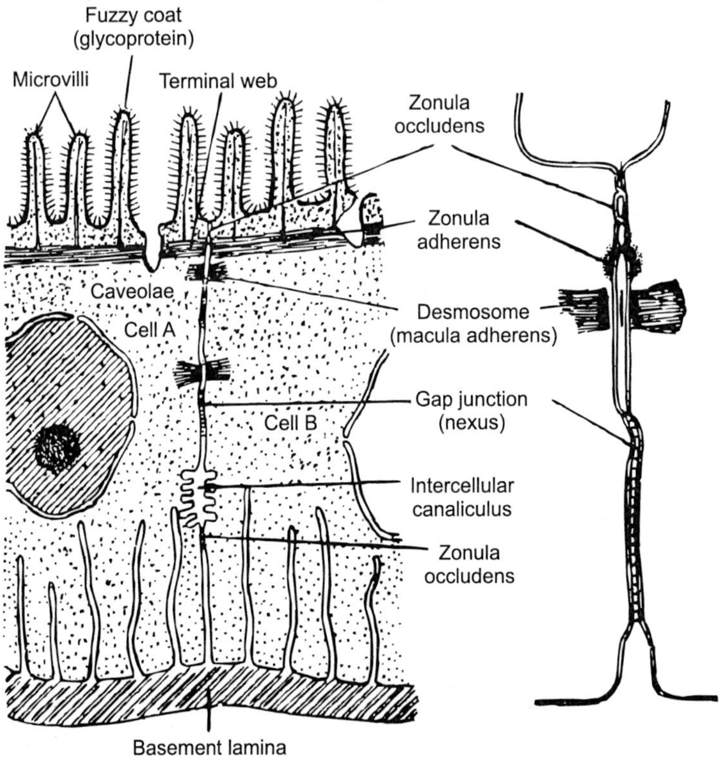

Fig. 3.15: Desmosome

Tonofilaments

Tonofilaments are also called keratin filaments. These are complex class of intermediate filaments which consists of prekeratin or cytokeratin filaments. These filaments anchor to the cell surface and tend to converge upon the desmosomes.

In man, there are as many as 19 different cytokeratins. Tonofilaments are not contractile. They form a structural framework for the cell cytoplasm. Mammalian cytokeratins are α-fibrous proteins that are synthesised in cells of the living layers of epidermis. They are composed of multiple polypeptides ranging in size between 47,000 to 58,000 daltons and have a three chain structural unit.

Cytoskeleton

The network of fibres found in the cell is known as cytoskeleton. The fibres of the cytoskeleton extend throughout the cell and form inter connections between the cell membrane and cell organelles. It is composed of fibrous proteins of the cytoplasm which help to maintain cell shape and to give contractibility to the cell. It also helps to facilitate communication among intracellular organelles. It also helps in cell locomotion or the movement of protoplasm, i.e. cyclosis.

Cytoskeleton consists of three components, the thickest tubular component called *microtubules*, thinnest fibres called *microfilaments* and fibres of intermediate thickness called *intermediate filaments*.

Types of Desmosomes:

There are three main types of desmosomes, namely belt desmosomes, spot desmosomes and hemidesmosomes.

(1) Belt Desmosomes (Zonula Adherans): These are formed at the interface between columnar cells just below the free surface. It is also called terminal bar or intermediary junction.

Under electron microscope, the belt desmosomes form a band that girdles the inner surface of the cell membrane. This band possess a web of 6 nm actin microfilaments and another group of interwoven intermediate filaments or tonofilaments of 10 nm. Actin filaments are contractile whereas intermediate filaments play a structural role.

(2) Spot Desmosomes (Macula Adherans): It is also called a fastening body or a binding body. These are disc or circular points of contact about 0.5 μm in diameter between the plasma membranes of adjacent cells. They are separated by a distance of 30-50 nm. The spot desmosomes consists of a disc shaped plaque (0.5 μm diameter) on the cytoplasmic surface of each cell membrane. Non-contractile tonofilaments from the structural framework of cytoplasm are attached to the plaques. Thin filaments called trans membrane linkers or desmocollins arising from the plaques fill the intercellular gap between the membranes.

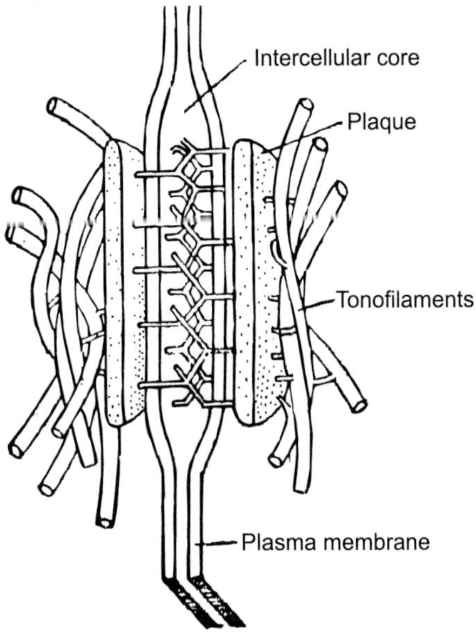

Fig. 3.16: Spot desmosome

They provide mechanical coupling. Sometimes a discontinuous middle dense line of coating material may be observed in the intracellular core. Cellular adhesion at the desmosomes depends mainly on this extracellular material.

(3) Hemidesmosomes: In the basal region of some epithelial cells, only half desmosomes may be present, called hemidesmosomes. They are useful as anchoring sites for bundles of tonofilaments and also for joining the cell membrane of epithelial cells to the underlying basement membrane.

(4) Septate Desmosomes: Wood (1959), **Lock and Gauranten** (1967) have reported different types of desmosomes in the epithelia of invertebrates and named them as septate desmosomes. In these desmosomes, the two plasma membrane remain separated by a distance of 150 Å - 200 Å and both remain joined by many transverse parallel septa. These septa are found to be continuous upto outer proteinous layers of plasma membranes. These desmosomes lack the intracellular cementing substance and tonofilaments of the desmosomes. They act as attachment or adhesive devices of cells and they also permit intercellular communication and electrical coupling.

(5) Terminal Bars (Zonula adherans): The terminal bars are also known as intermediary junctions or zonula adherans. The terminal bars are similar to desmosomes except that they lack tonofibrils. In terminal bars, the plasma membrane is thickened and the cytoplasm of thickened area is dense. The terminal bars occur in the intermediate portion of the plasma membrane of columnar cells just below the surface.

(6) Inter-digitations: At some places, the plasma membrane of two adjacent cells gives out finger like projections known as inter-digitations. The inter-digitations may become further complicated by the development of desmosomes, terminal bars, etc.

Classification of Desmosomes:

Desmosomes are classified in several ways:

(A) According to shape

(i) **Macular Desmosomes:** They are having spot like shape and are often called maculae adherans.

(ii) **Zonula Desmosomes:** Those which encircle a cell like a belt or zone and bind it to its neighbours all around its circumference at the level of the zonular desmosome is often called terminal bars.

(B) According to the structure to which cell surface is bound at desmosome

(i) **Antodesmosomes:** Those which bind an area of cell surface to another desmosome on another area of the surface of the same cell.

(ii) **Homodesmosomes:** Those that bind an area of cell surface of one cell to a desmosome on the cell surface of another cell. They may be either

Isodesmosomes: Binding one cell to another of the same type

Allodesmosomes: Binding one cell to another cell of a different type

(iii) **Heterodesmosomes or Hemidesmosomes:** Those which bind an area of cell surface to some non-cellular structure.

(C) According to similarity or dissimilarity of structure

(i) Symmetrical Desmosomes

(ii) Asymmetrical Desmosomes

(D) According to Function

(i) Adherent Desmosomes: Highly specialised for resisting tensile forces. They are all filamentous.

(ii) Occluding Desmosomes: Useful for sealing one cell membrane to the other and form barrier to diffusion and fluid exchanges between one extracellular fluid compartment and the other. e.g. septate desmosome.

(iii) Synaptic Desmosomes: Provide low resistance to the passage of ions between cytoplasm of two otherwise separated cells, joined by the desmosomes.

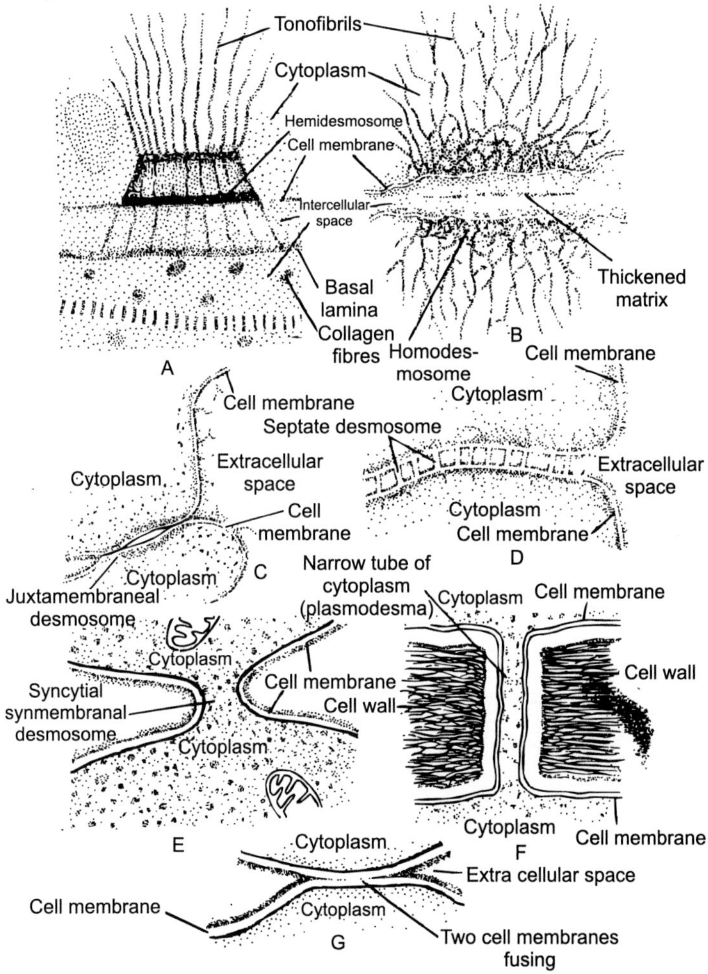

Fig. 3.17: Different Types of Desmosomes

(E) According to the relation between the membranes of two cell surface areas involved in desmosomes of a pair

(i) Dismembranal Desmosomes: The participating cell membranes are separated by an appreciable intercellular interval, which nevertheless, contains materials binding the two cell surface together.

(ii) Juxtamembranal Desmosomes: The cell membranes are very closely and intimately opposed in a parallel array but yet not fused with each other.

(iii) Synmembranal Desmosomes: The cell membranes are fused. It includes septate, syncytial and juxtalaminar desmosomes.

(F) According to the relation of the cytoplasm of the two participating cells

(i) Apocytical Desmosomes: Those in which the cytoplasm of the two participating cells are completely separated by membrane of high electrical resistance and are not electrically coupled.

(ii) Haptocytical Desmosomes: Cytoplasm of two cells is completely separated by membrane of low electrical resistance and is electrically coupled.

(iii) Syncitial Desmosomes: There is no membrane between two cells. Some plasmodesmata represent this type. These are also filamentous and non-filamentous desmosomes in which tonofibrils are present in former forms and tonofibrils are absent in later case.

3.5.4 Plasmodesmata

Sometimes the cells are connected by bridges of cytoplasm passing between pores of cell wall or plasma membrane between the adjacent cells, such connections are called *plasmodesmata* (singular-plasmodesma). They are usually simple but anastomosing plasmodesmata may also found. Their number and distribution vary considerably. Like gap junctions, plasmodesmata provide intercellular channels for molecules of about 1000 molecular weight including a variety of metabolic and signaling compounds. These structures are prominent in plant cells and depending on the plant type, the density of plasmodesmata varies from 1 to 10 per/μm^2.

Electron micrographs of plasmodesma show that it is a roughly cylindrical, membrane lined channel with a diameter of 20 to 60 nm and it traverses cell walls upto 90 nm thick. Plasmodesmata is a single narrow tube called desmotubule. They are prominent in plant cells but density varies in different plant types. The desmotubule is continuous with elements of endoplasmic reticulum membrane of each of the connected cells.

Between the outside of the desmotubules and the inner face of the cylindrical plasma membrane is an annulus of cytosol. It is constricted at each end of the plasmodesma. These constrictions may regulate the flux of molecules through the annulus that joins the two cytosols.

The plasmodesmata are needed in cell to cell communication. Many normal metabolic products such as sucrose are transported from cell to cell. As with gap junctions, movement of molecules through plasmodesmata is reversibly inhibited by an increase in cytosolic Ca^{2+}. Certain plant viruses and viroids can enlarge plasmodesmata in order to use this route to pass from cell to cell.

Through the plasmodesmata, cytoplasmic continuity is often maintained between adjacent cells. Through them, the material can pass from cell to cell.

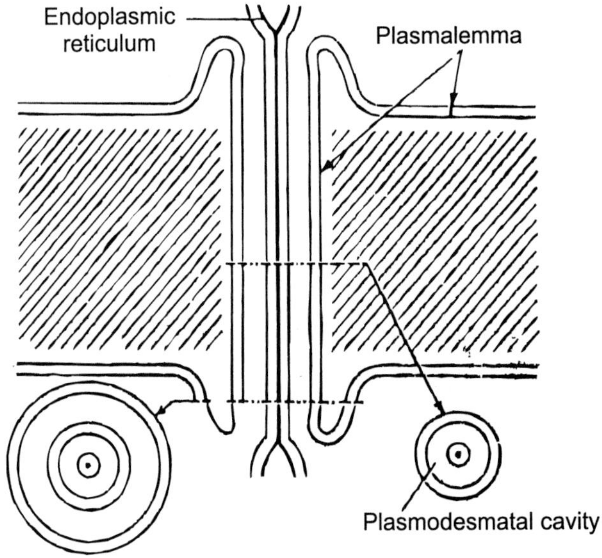

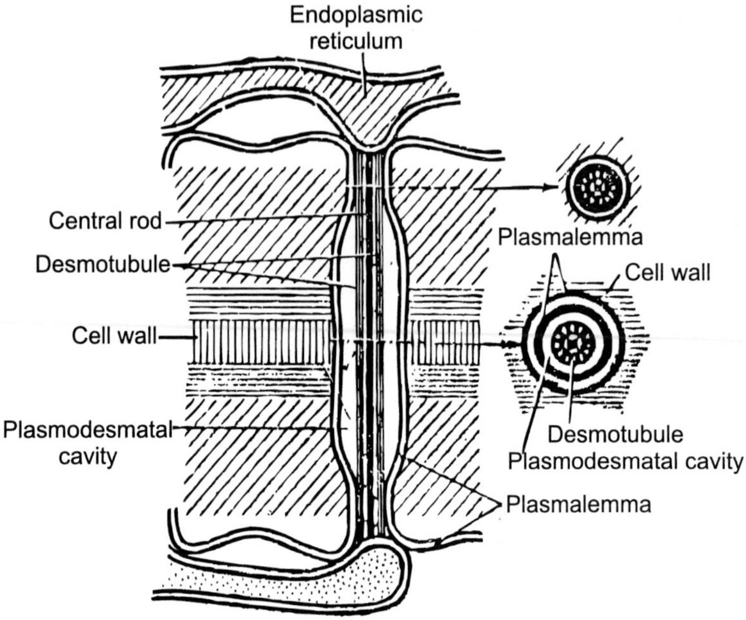

Fig. 3.18: Plasmodesmata

POINTS TO REMEMBER

- The plasma membrane regulates what enters and leaves the cell.
- The plasma membrane is made up of lipoproteins.
- The plasma membrane contains integral and peripheral proteins.
- Carbohydrates of plasma membrane are in the form of glycolipids and glycoproteins
- The fluid mosaic model proposed by Singer and Nicolson is the most accepted model for plasma membrane
- The narrow invaginations or finger like projections of plasma membrane at the cell surface are known as microvilli.
- Caveolae are cave like invaginations of the cell surface.
- Tight junctions are located at the distal region between the adjacent cells where actual fusion of the outer membrane of the adjoining cells occurs.
- Gap junctions are called as communicating junctions located at specialised region of contact between opposed plasma membranes of the adjacent cells.

- Desmosomes are abundant in cells of skin, heart muscle, and the neck of the uterus.
- Desmosomes form early in embryonic development and play an important role in maintaining cell position during development.
- Tonofilaments are filaments anchor to the cell surface and tend to converge upon the desmosomes.
- Cytoskeleton fibres extend throughout the cell and form inter connections between the cell membrane and cell organelles.
- Plasmodesmata are bridges of cytoplasm passing between pores of cell wall or plasma membrane between the adjacent cells.
- Transport of metabolites through plasma membrane takes place either by passive transport or active transport.
- The cell ingests or swallows foreign bodies, bacteria, harmful matter and other inert substances by a process called phagocytosis.

EXERCISE

1. What is the significance of plasma membrane?
2. Describe the ultra structure, chemical composition and functions of plasma membrane.
3. With a neat and labelled diagram explain the fluid mosaic model of plasma membrane.
4. Describe in brief functions of plasma membrane.
5. Explain mechanism of pinocytosis, phagocytosis and active transport through plasma membrane.
6. Give a brief account of fluid mosaic model of plasma membrane.
7. Explain difference between pinocytosis and phagocytosis.
8. Write short notes.
   a. Active transport
   b. Passive transport
   c. Pinocytosis
   d. Phagocytosis
   e. Fluid mosaic model
   f. Functions of plasma membrane
   g. Membrane receptors

   h. Microvilli
   i. Desmosomes

9. Define plasma membrane. Discuss its structure, chemistry and function.
10. Which is the most accepted model of plasma membrane? Add a note on functions of plasma membrane.
11. Define membrane receptors and add a note on the mobile hypothesis on fluidity of the membrane.
12. Describe the modifications of plasma membrane.
13. Differentiate the following:
   a. Active and Passive transport
   b. Pinocytosis and Phagocytosis
   c. Cell wall and Plasma membrane
   d. Gap junction and Tight junction
   e. Simple diffusion and Facilitated diffusion

❖❖❖

4

CHAPTER

NUCLEUS: STRUCTURE AND FUNCTION

CONTENTS

4.1 SIZE, SHAPE, NUMBER AND POSITION

Size:

The size of the nucleus also varies according to the cell type. But in general, a ratio exists between the nuclear volume and the volume of the cytoplasm, and this is characteristic for each cell type. This can be expressed numerically by the nucleoplasmic index. (NP)

$$NP = \frac{\text{Nuclear Volume (V)}}{\text{Volume of Cytoplasm } (V_c) - \text{Nuclear Volume } (V_n)}$$

Shape:

The shape of the nucleus also varies in different cells. It is generally spherical in cuboidal or polyhedral cells, and disc-shaped in squamous epithelium. In columnar epithelial cells, nucleus is ovoid and elongated. The nuclei are bilobed or multilobed in leukocytes. Horse-shoe shaped, branched, pyriform (kidney shaped) spindle shaped, elliptical nuclei have also been observed in other cells.

Number:

Almost all the higher cells possess nucleus. The mature red blood cells of mammals are without nuclei; however they are present in the immature stages of the RBCs. Most cells have a single nucleus, therefore they are called mononucleate cells, but there are many cells which possess more than one nucleus, and therefore they are called poly or multinucleate cells. If two nuclei are present the cell is called as a binucleate cell. Unicellular organisms like *Parameocium, Vorticella, Balantidium* are polynucleated cells. The polykaryocytes of bone marrow have up to 100 nuclei. Some cells do not show cellular boundaries in the mass of protoplasm. Such multinucleate mass of protoplasm is known as a *syncytium*.

Position:

The position of nucleus in a cell is variable. In embryonic cells, the nucleus is central in position but may later become displaced. In adipose tissue cells, nucleus is peripheral, whereas in glandular cells, nucleus is situated at the basal region.

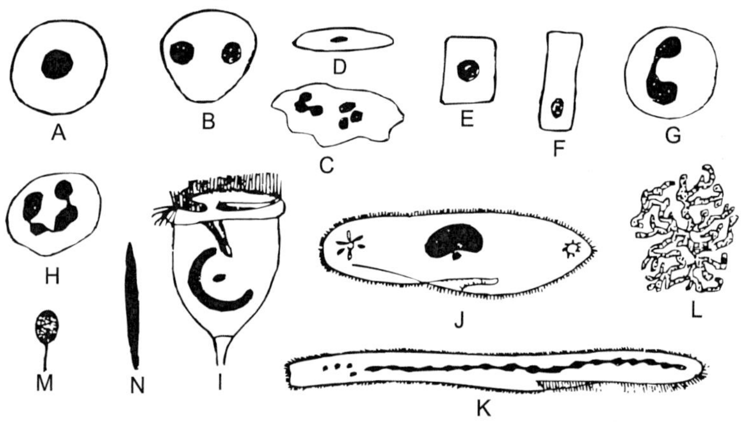

Fig. 4.1: Number and shape of nuclei

(A)	Mononucleate cell	(H)	Multilobed nucleus
(B)	Binucleate cell	(I)	C-shaped nucleus of *Vorticella*
(C)	Polynucleate cell	(J)	Pyriform nucleus of *Parameocium*
(D)	Disc Shaped nucleus	(K)	Moniliform nucleus of *Spirostomum*
(E)	Spherical nucleus	(L)	Labyrinthine nucleus of *Platyphylla*
(F)	Oval nucleus	(M)	Elliptical nucleus of sperm
(G)	Bilobed nucleus	(N)	Spindle shaped nucleus of sperm

4.2 ULTRASTRUCTURE OF NUCLEAR MEMBRANE AND PORE COMPLEX

The nucleus is bounded by the nuclear envelope or membrane or karyotheca. Within the nucleus is a clear or slightly acidophilic mass called the nuclear sap or nuclear matrix. In the interphasic stage, the nuclear sap contains twisted filaments of chromatin which are called chromosomes. False nucleoli or karyosomes are nothing but flakes of condensed regions of chromatin found in the nucleus. One or more spherical bodies called nucleoli or plasmasomes are also observed. Thus, in the nucleus following structures are present.

(1) Nuclear envelope or membrane

(2) Nuclear sap or nuclear matrix

(3) Chromatin

(4) Nucleoli

4.2.1 Nuclear Envelope

The nucleus is separated from the cytoplasm by a double membrane called the nuclear envelope. The two membranes are separated from each other by a perinuclear space of varying width. Each layer is about 100 to 300 Å apart, leaving a discontinuous perinuclear space that communicates with the channels of cytoplasmic membrane. The envelope is absent for a very brief period only at the time of cell division.

Structure of the Envelope:

The two membranes of the nuclear envelope are roughly parallel, and each membrane is 70-80 Å thick. **Outer membrane** has connections with the endoplasmic reticulum and has ribosomes on the outer side. The **inner membrane** possesses a crystalline layer, often coated with filaments and fibrous structures, some of which extend deeper in the nucleus and others may attach to the chromatin material. The nuclear membranes are separated by **perinuclear space** which is discontinuous except in some areas where the membranes join to form **pore complex**. The two membranes around these pores are in continuity and form a rounded lip around the circular area. The space between the two membranes is irregular, discontinuous and has varying width. The perinuclear space is filled with fluid communicating with the endoplasmic reticulum.

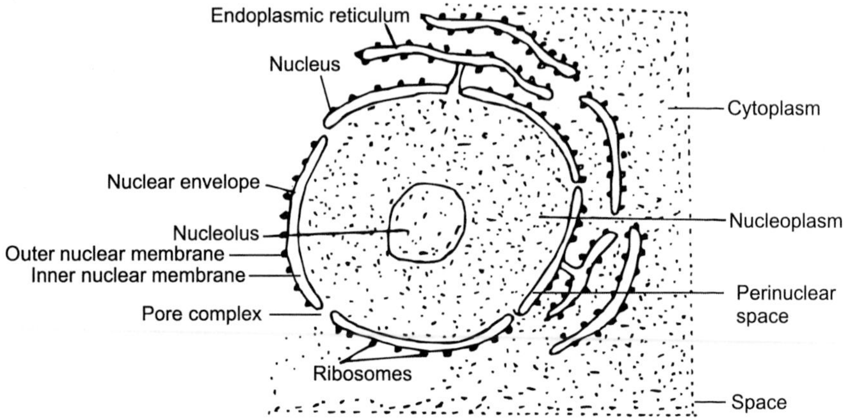

Fig. 4.2: Nucleus with nuclear envelope and nuclear pores

4.2.2 Nuclear Pore Complex **(April 2014)**

The nuclear envelope is interrupted at intervals by nuclear pores which act as passage ways for transport of various types of materials. In plant cells, these pores are irregularly and sparsely distributed over the surface of the nucleus, but in certain animal cells (e.g. amphibian oocyte) the pores are numerous and regularly arranged. It has been observed that the number of pores per unit area of the envelope varies with cell types and with the physiological state of the cell. The pores are aligned with the nucleoplasmic channels. The number of pores range from 100 to 5×10^7 in different cells. It seems that the pore number is related to pore density and the number of RNA molecules being transported from the nucleus.

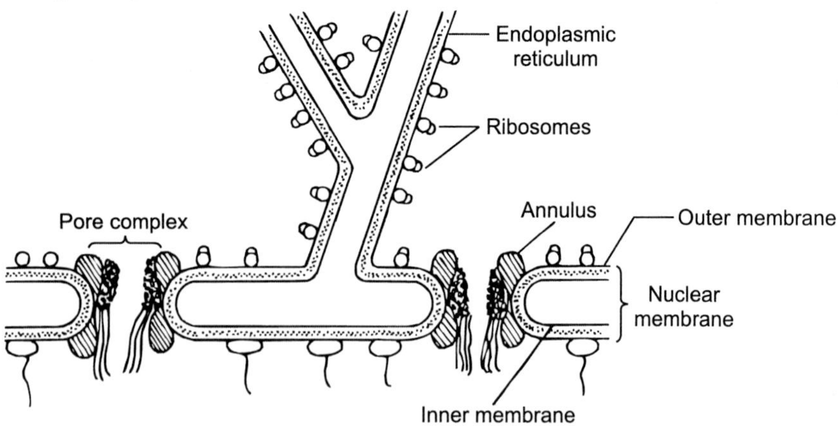

Fig. 4.3: Nuclear envelope and its relationship with endoplasmic reticulum

The pores are also arranged in rows or hexagonal clusters. Electron microscopic studies revealed that the pores contain some electron-dense materials and are enclosed by some annular material, which is granular. These circular structures are called annuli, which along with pores from a pore complex. Pore diameter also varies from cell to cell, generally from 50 to 100 nm. The annulus is ring like and consists of eight granules, each with a diameter of 15 – 17 nm. The matrix of the annulus may be amorphous or fibrillar which is digested by trypsin and hence is proteinaceous in nature. It is however, not hydrolysed by nucleases. The pore complex is a rigid structure.

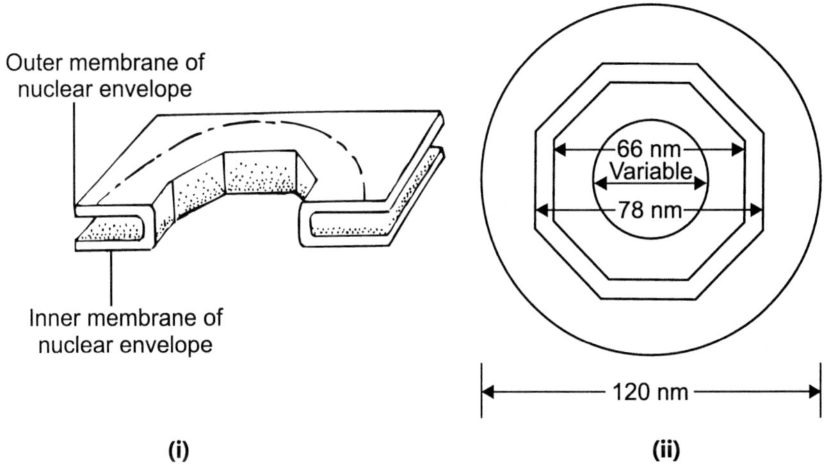

Fig. 4.4 (a): Structure of nuclear pore, diameter of nuclear pore

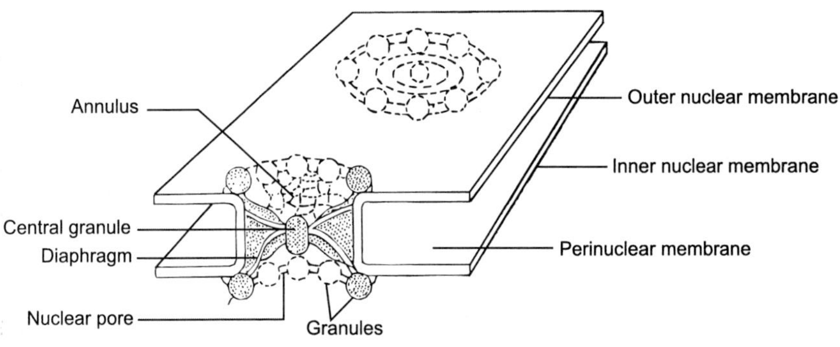

Fig. 4.4 (b): Nuclear pore complex in the nuclear membrane (Frank's model)

Below the inner membrane of the nuclear envelope, there is an internal dense lamella, composed of homogeneous material called the honey comb layer, which is bordered by lumps of chromatin on the inner surface. The thickness of the lamella varies from 200 to 800 Å. The nuclear envelope shows connections with the nuclear components as well as the cytoplasmic region. The dense chromatin region or layer is interrupted below the pores, giving rise to intranuclear channels, which open on the nuclear surface through the pores.

4.2.3 Nuclear Sap/Nuclear Matrix

The body of the nucleus within the nuclear envelope is made up of a dense jelly-like mass which is composed of two elements, each of which is dynamic in nature and has complex chemical composition. These elements of nuclear sap are the nuclear gel or karyolymph and the chromatin.

The nuclear gel: It is highly granular, containing fibrous material. It is rich in proteins, which account for more than 90% material. A small percentage of DNA, RNA and phospholipids is also present. The proteins are largely acidic and they have an unusual property of expanding and contracting under the influence of Ca^{2+} and Mg^{2+} ions. In this region, events like replication of DNA, transcription and transport of substances take place.

The Chromatin: The other components of the nuclear matrix include nucleolus and chromatin. The chromatin is present in a condensed form called heterochromatin, which is present in patches near the nuclear envelope, either scattered in the interior or surrounding the nucleoli. The matrix also has a network of non-chromatin material forming the interchromatin matrix.

It was first thought that the site of DNA replication lies in the vicinity of nuclear envelope. This assumption was based on the observation that most of the chromatin material is localised on the nuclear side of the envelope. However, latest findings indicate that the chromatin is suspended in the matrix. Matrix is made up of some proteins which can be easily digested or hydrolysed by enzyme pepsin. Nuclear matrix is also the site of transcription because a certain amount of residual RNA is found associated with the matrix protein. Subsequently, RNA undergoes post transcriptional changes in the matrix itself where it is processed to smaller RNA pieces, which ultimately pass through the nuclear pores into the cytosol.

Functions:

The nuclear matrix plays three important functions such as:

(i) DNA replication

(ii) DNA transcription

(iii) Post transcriptional processing

4.2.4 Nucleolus **(April 2014)**

The nucleolus was first recognised by **Fontona** in 1874. It is generally spherical shaped body situated in the nucleus either in a central or peripheral region. The number of nucleoli in each nucleus depends on the species or the sets of chromosomes present in the nucleus. The number of nucleoli in each nucleus may be 1, 2, or 4. Nucleolus is the characteristic feature of eukaryotic cells, while it is absent in prokaryotic cells. The size of nucleolus depends on the synthetic activity of the cell. Thus, the cells with little or no synthetic activities e.g. sperm cells, muscle cells, are found to contain smaller or no nucleoli, while the oocytes, neurons and secretary cells which synthesise proteins or other substances contain large sized nucleoli. The size varies from 1 to 5 μm in diameter.

General Organisation:

Under light microscope, the nucleolus appears as a fluid or semisolid homogeneous body. It shows four main components namely granules, fibrils, matrix and chromatin.

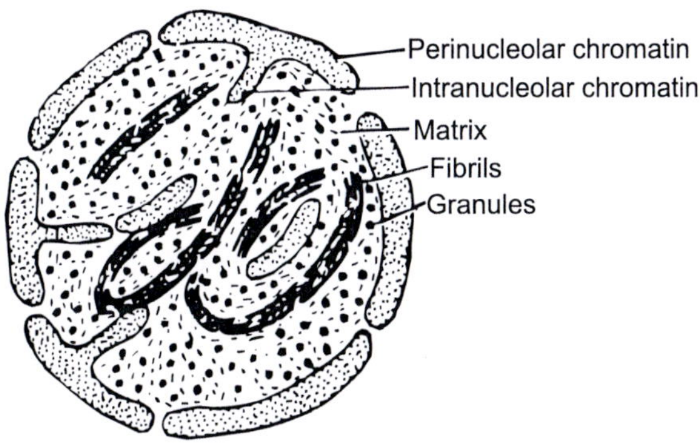

Fig. 4.5: Nucleolus

(i) Granules: The granular region generally occurs densely at the periphery of the nucleolus. These granules are of 150 to 200 Å in diameter. The granules contain ribo-nucleoproteins in which RNA and proteins are in the ratio of 1 : 2. These granules are generally modified into ribosomes. All the granules are connected together by a thin filament resembling a string with beads.

(ii) Fibrils: Fibrillar region is also known as *nucleolonema*. It is mainly composed by many fine fibres called fibrils. They are made up of ribonucleic proteins which are 50 – 80 Å in length. Both granules and fibrils are digested by ribonuclease enzyme.

(iii) Matrix: This is also called amorphous or proteinous region. The presence of granules and fibrils point to the fact that the nuclear matrix is the major site where DNA replication takes place first and later it gets associated with the nuclear envelope.

Nucleolus also contains electron dense chromatin material, usually at the periphery. Chromatin is composed of fibres of 100 Å thickness. The chromatin contains DNA, which serves are template for RNA synthesis. Surrounding the nucleolus, is the perinucelolar chromatin. Projecting into the nucleolus from the perinucleolar chromatin are septa like trabeculae which constitute intranucleolar chromatin.

Chemical Composition:

As far as the chemical composition of nucelolus is concerned 5 to 10% of the nucleolus is made up of RNA and remaining part is made up of protein. The main protein components are phosphoproteins. Acid phosphatase, nucleoside phosphorylase and NAD^+ synthesising enzymes have been localised in the nucleolus. It is surrounded by a ring of Feulgen positive chromatin which actually represents the heterochromatin regions of chromosomes.

Functions of Nucleolus:

(1) **Ribosome Formation:** Nucleolus is involved in the biogenesis of ribosomes.

(2) **RNA Production:** Nucleolus is one of the most active sites of RNA synthesis. It produces about 70-90% of cellular RNA in many cells. It is also the source of ribosomal RNA (rRNA).

(3) **Protein Synthesis:** Maggio (1960) and others have suggested that the protein synthesis takes place in the nucleolus. Some cell biologists suggested that ribosomal proteins are synthesised by the nucleolus.

(4) **Role in Mitosis:** The nucleolus plays a significant role in mitosis.

4.2.5 Nucleocytoplasmic Interactions

Most of the nucelic acids like DNA and RNA of a cell are formed in the nucleus and those found in the cytoplasm are thought to have migrated there. The exceptions to this rule are the DNA of some cytoplasmic organelles. Messenger RNA is formed in the nucleus, in association with the chromosomes. Ribosomal RNA is also formed in the nucleus, especially in the nucleolus. Since, some protein synthesis goes on in the nucleus, m-RNA and ribosomes first become associated with the nucleus, to form functional polysomes. Most polysomes are found in the cytoplasm and it is assumed that they migrate through the pores in the nuclear membrane. Recently, evidence has been obtained for the existence of a particle which appears in the nucleus but rapidly becomes associated with polysomes in the cytoplasm. These particles contain a precursor of the 40S ribosomal sub-particle and possibly mRNA. It may associate with 60S sub-particle forming one molecule of mRNA. Other ribosomes may then become attached and may be extensively reused in the cytoplasm.

Some polysomes remain free in the cell sap but become bound to the membranes of the ER. Cytoplasmic nucleic acids are synthesised in the nucleus. RNA might act as a template for synthesis of other RNA molecules in the cytoplasm. The cytoplasm might have DNA containing structures. Cytoplasmic DNA falls into three categories. In some amphibian eggs, large amount of DNA is present in the cytoplasm. It may function in regulating the early development of the embryo. The cells of microorganisms and protozoa have self replicating DNA containing particles called episomes. They exist free in the cytoplasm or in intimate association with the chromosome.

Cellular activity is controlled by the nucleus and more specifically by DNA located in the chromosomes. The RNA involved in protein synthesis originates in the nucleus and is then transferred to the cytoplasm. It is generally accepted that some macromolecules, small organic molecules, water and ions can pass across the nuclear envelope. These macromolecules can cross the nuclear envelope through nuclear pores by active transport, by blebbing (an irregular bulge in the plasma membrane of cell caused by localised decoupling of cytoskeleton of plasma membrane) and exchange through the endoplasmic reticulum. Many studies suggest that granular material of nucleolar origin pass from the nucleus to the cytoplasm, and may be involved in the biogenesis of ribosomes.

4.3 CHROMATIN

Euchromatin and Heterochromatin

The chromosomes are composed of a substance known as chromatin material (DNA + Nucleoproteins, etc.) which occur in two forms i.e. ***euchromatin*** and ***heterochromatin***. *Heitz* (1928) described heterochromatin as those regions of a chromosome that remain condensed during interphase and early prophase. The rest of the chromosome remains in a non-condensed state and is called as euchromatin. Heterochromatin remains present in close contact with the nucleolus and is responsible to the nucleolar organiser.

During mitosis, the heterochromatic regions appear localised at the centromeres. Such regions may stain more strongly or weakly than the euchromatic region.

Fig. 4.6: Structure of a typical chromosome showing euchromatin and heterochromatin

Electron microscopic studies reveal that, heterochromatin and euchromatin differ in their physical structure. Heterochromatin is composed of 250 Å fibrils, whereas euchromatin contains 30 to 80 Å fibres; so the heterochromatin represents a condensed intercoiled state of chromatin, containing two to three times more DNA than euchromatin. However, it is genetically inert because it is not transcribed by RNA (i.e. transcription) and protein.

POINTS TO REMEMBER

- Cells that have a single nucleus are called mononucleate cells while those that possess more than one nucleus are called poly or multinucleate cells.
- The nucleoplasmic index - a ratio of the nuclear volume and the volume of the cytoplasm is characteristic for each cell type.
- The nucleus is separated from the cytoplasm by a double membrane called the nuclear envelope.
- The two membranes of the nuclear envelope – outer and inner are roughly parallel.
- The nuclear pores present in the nuclear envelope are passage ways for transport of various types of materials.
- The nuclear gel and the chromatin are the elements of nuclear sap.
- Nucleolus is a fluid or semisolid homogeneous body having four main components - granules, fibrils, matrix and chromatin.

EXERCISE

1. Describe the structure of interphase nucleus.
2. Discuss the structure and functions of nucleus.
3. Name the various components of nucleus. Give an account of the structure and functions of each.
4. Describe morphology of nucleus and functions of nuclear envelope.
5. Describe the morphology, position and size of nucleus and add a note on functions of nucleus.
6. Write short notes on:
   a. Nuclear envelope
   b. Functions of nucleus **(April 2014)**
   c. Nuclear matrix **(April 2014)**
   d. Nuclear cytoplasmic interactions
   e. Functions of nuclear envelope
   f. Pore complex
   g. Nucleolus
   h. Nucleoprotein

❖❖❖

5

CHAPTER

ENDOMEMBRANE SYSTEM

CONTENTS

5.1 ENDOPLASMIC RETICULUM

The endoplasmic reticulum is found in almost all animal and plant cells except erythrocytes and prokaryotes. It is an interconnected network of membrane vesicles. ER was first observed by **Garnier** but its ultrastructure was first given by **Porter, Claude, and Fullam**. The term "Endoplasmic Reticulum" was given by Porter. He stated that "the endoplasmic reticulum is a finely divided vascular system, extending from the nucleus throughout the cytoplasm to the margin of the cell".

5.1.1 Morphology

The endoplasmic reticulum comprises of a complex system of cisternae, tubules and vesicles. The cisternae are broad, flattened, membrane bound spaces about 40 to 50 mμ thick and often arranged in parallel stacks. They are interconnected with each other. Tubules are more diverse in their shapes than cisternae. Their diameter ranges from 50 to 100 mm. Vesicles are rounded in shape and vary from 20 to 500 μm in diameter. Three forms of endoplasmic reticulum may appear in a single cell at the same time or may appear at different times during the cell cycle.

5.1.2 Types of Endoplasmic Reticulum

There are two basic morphological types of the endoplasmic reticulum (1) the rough endoplasmic reticulum or granular form and (2) the smooth endoplasmic reticulum or agranular form. The rough endoplasmic reticulum is so called because the membranes are covered with ribosomes, which give them a rough appearance. The smooth endoplasmic reticulum membranes are not covered with ribosomes.

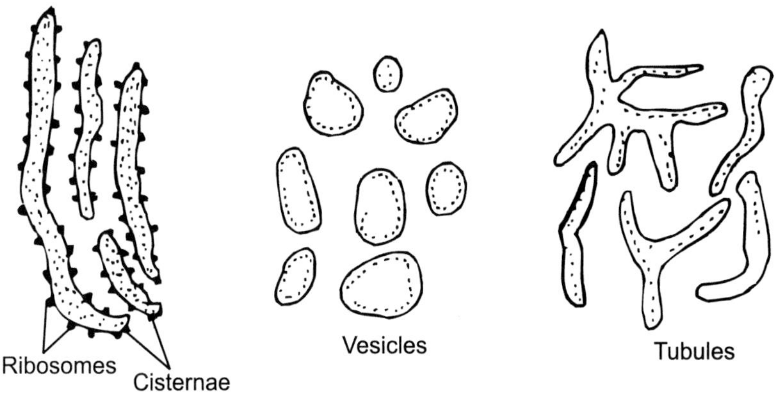

Fig. 5.1: Types of the endoplasmic reticulum

The rough endoplasmic reticulum (RER) are extensively developed in the cells which actively synthesise proteins, e.g. enzymes secreting cells, while smooth endoplasmic reticulum (SER) is characteristic of cells in which synthesis of non-protein substances like phospholipids, glycolipids, and steroid hormone takes place, e.g. adipose tissue cells, adrenocortical cells and interstitial cells of the testis.

The smooth ER is less stable than rough ER. The granular ER is involved in protein synthesis but the agranular ER has many functions other than protein synthesis. A large rough ER is indicative of metabolically active cell. The ER membrane divides the cell into atleast two compartments. Secretory products, pinocytic material, phagocytic material are found in the compartment enclosed by the membranes.

A modified form of smooth ER is the sarcoplasmic reticulum found in striated muscles. The myeloid body in the pigment cells of the retina of the frog is modified smooth ER. It consists of a number of tubules packed over one another. Myeloid bodies are light-sensitive regions in the cell and play an important role in photoreception.

Microsomes are not the normal structural units of the cells, but represent a functional unit. When cells are ground or homogenized, the cell membrane breaks into fragments. The fragments then round off to bodies called microsomes. They are generally the fragments of ER. Under electron microscope they appear as a membrane bound vesicles, 500 to 1000 Å in diameter. This fraction contains most of the RNA (50 to 60% RNA of the cell) and phospholipids of the cell and shows high rate of protein synthesis than in non-ribosomal endoplasmic reticulum. The microsome is a heterogeneous fraction containing not only the endoplasmic reticulum, but also containing Golgi membrane, ruptured plasma membrane and other cell fragments.

The lipid content of microsomes is estimated to be 30-50% of which about 70% is phospholipids; 50-90% of the phospholipid content is in the form of lecithin and cephalin. The ER membrane has the typical three layered unit-membrane structure in some region, while in other region; it may show a globular (micellar) structure. It thus appears to have a combination of triple layer and micellar structures.

5.1.3 Functions of Endoplasmic Reticulum

1. Mechanical Support: The endoplasmic reticulum divides the fluid content of cell into compartments and thus provides additional mechanical support for the colloidal structure of the cytoplasm. These compartments make possible the existence of ionic gradients and electrical potentials along the ER membranes.

2. Cellular Metabolism: The membrane of the reticulum provides an increased surface for metabolic activities within the cytoplasm, such as synthesis of cholesterol, triglycerides and other lipids, indicating their probable role in lipid metabolism.

3. Protein Synthesis: Proteins are synthesised on the ribosomes and enter ER cisternae through channels in the membrane. ER facilitates protein folding and the transport of synthesised proteins in sacs. It has been pointed out that the free ribosomes are not as active in protein synthesis as are those associated with ER. This suggests that the membrane contributes in some way to the efficiency of the process.

In pancreatic aciner cells of the dog, dense granules are seen in the cisternae. In the liver cell of salamander *Bairachoseps attenuatus* protein crystals have been found in the granular cisternae. The ER thus has a property similar to that of the Golgi complex, namely concentration of products into dense granules.

4. Intracellular Transport: Protein, including enzymes, lipids and other cellular materials are transported and distributed to the various parts of the cell through ER. In some cases, these materials may accumulate and may be stored within the ER for considerable time.

5. Detoxification: Smooth ER is involved in the detoxification of many endogenous and exogenous compounds. Prolonged administration of certain drugs (phenobarbital) results in the increased activity of enzymes related to detoxification, as well as other enzymes and a considerable hypertrophy of the smooth ER.

6. Membrane Flow: Transport of ions, molecules and particles into and out of the cells may also take place through membrane flow. Thus substances like RNA and nuclear proteins may pass out from the nucleus to the outside of the cell by the following route:

Nuclear membrane → Pores → ER → Golgi complex → Plasma membrane → Outside. Particles enter the cell by a reverse membrane involving pinocytosis/phagocytosis by the plasma membrane and vesicle formation i.e. endocytosis.

7. Ionic Gradients: The sarcoplasmic reticulum of striated muscle is involved in increase in the concentration of calcium ions by an energy utilising ATP. The calcium ions are stored in the sarcoplasmic reticulum. When the muscle is stimulated by nerve impulses, hormone or other means, the calcium ions are released, leading to muscular contraction.

8. Formation of Plasmodesmata: Electron microscopic studies suggest that the endoplasmic reticulum in plants play a special role in the interconnection of cells through the cytoplasmic strands called plasmodesmata.

9. Formation of Secondary Wall: In the formation of secondary cell wall in plants, certain enzymes and metabolites may be carried by the reticulum to the region of wall synthesis.

10. Glycogen Synthesis and Storage: Residual glycogen has been found associated with the endoplasmic reticulum in fasting animals. On recommencement of feeding, there is an increase in the endoplasmic reticulum glycogen.

The enzyme *glucose-6-phosphatase* has been reported in the ER of rat liver cells. It has been linked to breakdown of glycogen (glycogenolysis). It has been also suggested that the SER membranes play a role in the synthesis of glycogen (glycogenesis).

11. Lipid Synthesis and Storage: Electron microscopic and autoradiographic studies of Stein and Stein (1967) suggested that the ER was the site of triglyceride formation. They injected radioactive labelled material (3H-palmitate and 3H-glycogen) into fasted ethanol treated rats. It was found that the labelled material was seen in the SER and RER after two minutes. Claude (1970) has suggested that triglyceride synthesis may be limited to the SER, while protein and perhaps phospholipids are synthesised in the RER.

The SER membrane also appears to be involved in the formation of lipoprotein complexes. The SER is also implicated in the initial stages of the breakdown of fatty acids. Conversion of fatty acid to acetyl coenzyme A esters occurs mainly in the ER.

The ER has been associated with the storage of lipids. When starved rats were fed with corn oil, the oil droplets first appeared in pinocytic vesicles, below the microvilli of intestinal epithelial cells. After about an hour, the lipid was found in the ER and the Golgi complex. Thus, the ER appears to have a general role in the intracellular and storage of lipids.

12. Synthesis of Cholesterol and Steroid Hormones: Cholesterol is an important precursor of steroid hormones. The major site of cholesterol synthesis is the ER. In rats, injected with radioactively labelled acetate, nearly all the labelled cholesterol is found in the microsomal fraction. The microsomes have also been shown to be the principal sites of cholesterol forming enzymes. The intermediates *squalene* and *lanosterol* in cholesterol synthesis are bound to the ER in rat liver cells. In liver cells, the SER is believed to be concerned with both the synthesis and storage of cholesterol.

In the testis, ovary and the adrenal cortex, the SER has a role in the synthesis of steroid hormones. A well developed SER has been demonstrated in the interstitial cells of the testis of the opossum and the guinea pig. The enzyme catalysing biosynthesis of androgens have been located in the SER. There is a strong correlation between the amounts of SER in cells and the capacity to synthesise steroid hormones.

13. Formation of Microbodies: Closely related with the ER, are microbodies, which are small granular bodies filled with an electron dense substance and limited by a single membrane. They have been found in the protozoa, yeast, higher plants, liver and kidney. In rat liver cells there are 70 - 100 microbodies with an average diameter of 0.6 to 0.7 µm. Microbodies are formed as dilations of the ER and frequently show connection with the ER cisternae. They are rich in enzyme peroxidase (and are hence also called peroxisomes), catalase, and D-amino acid oxidase. In plant cells, the enzymatic content is different, and the bodies are glyoxysomes because they include enzymes of the glyoxylate cycle.

Peroxisomes produce oxidative enzymes which generate hydrogen peroxide (H_2O_2). The enzymes are *urate oxidase, D-amino oxidase* and *hydroxylic acid oxidase.* They produce hydrogen peroxide which is toxic to the cell. Another enzyme produced is *catalase* which destroys hydrogen peroxide, and therefore probably plays a protective role. Peroxide shows a relationship with steroid synthesis and generation of NAD and NADPH.

14. Origin of Cell Organelles: There are evidences that show ER is the place for the origin of the organelles known as primary lysosomes.

15. Amphibian Development: There are evidences to suggest that the ER contributes in several ways to the development of the amphibian embryo.

16. Cell Differentiation: The ER is important in the process of cell differentiation and coordinating the differentiation.

5.2 GOLGI COMPLEX

The Golgi complex is supposed to be originated from the endoplasmic reticulum. They are a system of roughly parallel interconnecting flattened sacs situated close to the endoplasmic reticulum but physiologically separate from them. Golgi complex was first discovered by Golgi, an Italian neurologist in 1898 who named it "natural reticular apparatus". Various names have been given to this organelle by various workers. **Baker** (1951, 1953) referred to the Golgi complex as lipochondria because of the presumed lipid content. Structures similar to the Golgi complex have been found in plants. Botanists refer to them as ***dictyosomes.***

The term Golgi complex has been preferred because of the following reasons:

1. Electron microscopy demonstrates fairly discrete groups of laminar units.
2. Perroncito had earlier established a definite use for dictyosomes.
3. Each dictyosome appears to function in membrane building, product synthesis and assembly, and in the evolution of products as an individual organelle.

5.2.1 Morphology

The shape of Golgi is quite variable in somatic animal cells; even in the same cell, there is variation with functional stages. In some cases, it occurs as a dense reticulum of anastomosing trabeculae while in others as an irregular lenestrated plaque or hollow spheres united together. In nerve cells, it occurs as a reticulum of wide meshes around the nucleus.

The size of the Golgi is also variable. It is small in muscle cells but quite large in the nerve cells and gland cells.

The number of Golgi apparatus per cell is also variable. Some cells have been recorded as having a single apparatus; other cells with dispersed Golgi apparatus may have hundreds. Some glandular cells have only a single apparatus but of appreciable extent.

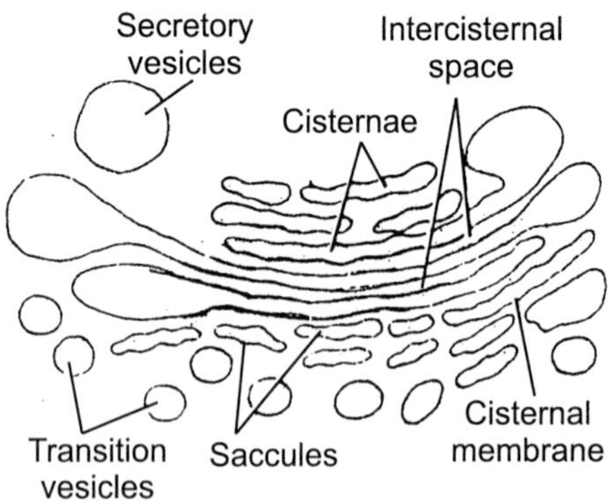

Fig. 5.2 : (a) Ultrastructure of Golgi Complex

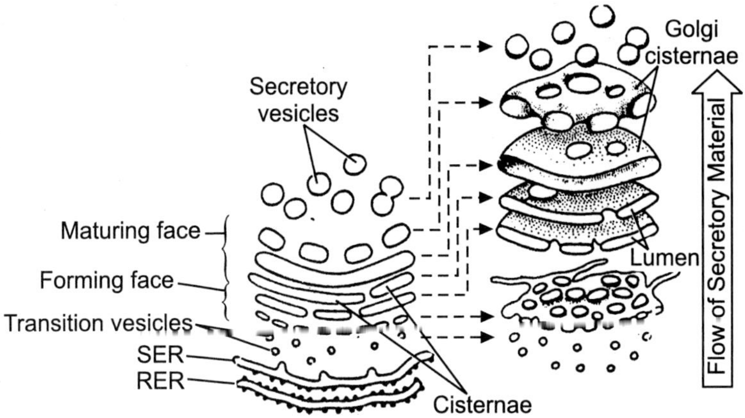

Fig. 5.2: (b) Diagram showing polarity and membrane flow in Golgi cisternae

The position of Golgi is relatively fixed for each cell type. In the cells which are ectodermal in origin, the Golgi is polarised and is present between the nucleus and the periphery of the cell. In the secondary exocrine gland, it is generally found between the nucleus and the excretory pore. In the endocrine gland cells, its polarity is variable except in thyroid, where it is oriented towards the centre of the follicle.

5.2.2 Structure of Golgi Complex

The first electron micrograph of Golgi complex was proposed by **Dalton and Felix**, 1954. They described the Golgi complex in the epididymis of the rat as consisting of three components; flattened sacs or cisternae, large vacuoles and small granules. These are all membranous structures and are characterised by the absence of ribosomes i.e. they are smooth membranes.

Morphologically Golgi complex lies between the ER and plasma membrane, the three membranes being linked by transitional structures and by secretory granules. This suggests that the Golgi complex membrane is in the intermediate state of differentiation between ER membrane and the plasma membrane. Generally, it consists of about 60% protein and 40% lipid; some proteins being common to those found in ER, but having fewer protein bonds. Very low level of RNA, DNA and polysaccharides has been noted in some Golgi functions. The natural lipid fraction contains mainly cholesterol, cholesterol esters and triglycerides. The enzyme concentrated in the Golgi fraction comprises thiamine pyrophosphatase and glycosyl transferase.

The cisternae or lamellae are the most constant elements of the Golgi complex. They consist of flattened; parallel sacs situated one upon the other to form stacks. The number of cisternae in a stack varies from species to species and some times from one developmental stage to another. The number of cisternae in a stack is about 4 to 5 in most animal and plant cells. In *Euglena*, the number may go up to 20. The membrane of cisternae is roughly 70 Å in thickness which encloses a cavity of about 150 Å wide whose edges are often dilated.

In certain plant cells the Golgi complex is polarised and has two well defined faces "forming face and maturing face". The forming or proximal face is the outer side, while the maturing or distal face is on the opposite side. It is also possible that a lamella of the ER buds off vesicles, which then become arranged on the forming face of the ER and may be converted into a Golgi lamella by loss of ribosomes. On the other side, is a maturing or secreting face, where the lamella buds off secretory vesicles. Thus, new lamellae are formed on the forming face and maturing lamellae are lost on the maturing face. The cisternae lie parallel to each other and are separated from each other by a space at about 200 to 300 Å. Chemical composition shows that, Golgi complex membrane is intermediate between those of ER and the plasmalemma. The membranes at the forming face of the Golgi complex are similar to the ER membrane and those of the maturing face, to the plasmalemma. The membranes of the Golgi complex are in dynamic equilibrium. They are continually receiving lamellae through budding off of vesicles from the smooth ER and losing membranes through formation of secretory vesicles.

In a few cases, dense material has been found between the cisternae of the Golgi complex and has been called intercisternal material. The composition and function of the intercisternal material is not known. It has been suggested that it acts as a cementing substance, holding the cisternae together.

Vesicles

The small vesicles are 400 to 800 Å in diameter. They are intimately associated with the cisternae and may show continuity with them. The small vesicles arise from the cisternae by budding or pinching off.

The large vacuoles are clear and generally lie at the edge of the Golgi complex. They represent modified and expanded cisternae in which the two membranes have become widely separated and the vacuolar space enlarged. They contain clear material which reduces metallic salts.

5.2.3 Functions of Golgi Complex

1. Golgi and Cell Secretion: There are definite evidences which suggest relationship between Golgi and cell secretion which was postulated by **Cajal**, 1914. The very fact that, the Golgi are very well developed in all the cells exhibiting high secretory activity, clearly suggest that they play main role in the formation of secretion granules.

Many electron microscope studies show that during vitellogenesis in oocytes, yolk products first become visible in the cisternae of Golgi complex. However, in these studies it is not clear whether the yolk actually synthesises the Golgi complex or only packs them. In certain oocytes, the Golgi complex appears to give rise to critical granules.

In some cells, formed products appear in the expanded end of the Golgi lamellae. In others, secretory products completely fill the cisternae. The secretory products within the Golgi lamellae are very similar to the contents of the secretory granules near the Golgi complex. In some cases the ends of the Golgi cisternae may be pinched off to form small secretory granules. These may then fuse to form large granules. In other cases, the individual cisternae on the "mature face" may be completely filled with secretory products and then become rounded to form secretory granules. New cisternae would then apparently be formed on the forming face. It is possible that the ER may become Golgi lamellae by loss of ribosomes.

2. Role in Protein Secretion: In pancreatic exocrine cells, the distribution of secretory proteins (digestive enzymes) has been investigated by Siekevitz and Palade in 1962.

(a) Proteins are formed on the ribosomes attached to the ER.

(b) These nascent proteins are then transferred into the ER.

(c) From here, they go to the Golgi complex. There is morphological evidence that the ER may bud off vesicles which travel towards the Golgi complex. The vesicle may then fuse with the membrane of the Golgi cisternae and release their content into the latter.

(d) In the Golgi complex, the proteins are concentrated and transformed into zymogen granules. The zymogen of the granules consists of enzyme precursor of the pancreatic juice. The membrane of the Golgi vacuole becomes the limiting membrane of the zymogen granules.

(e) The zymogen granules released from the Golgi complex, migrate to the surface of the cell. Here the limiting membrane of zymogen granules fuses with the plasmalemma, thus discharging its contents. This process is a form of "reversed pinocytosis".

Thus, it is apparent that the Golgi complex acts as a condensation for proteins formed on the ribosomes. The concentrated material is then packed for secretion.

3. Synthesis of Carbohydrates: Golgi also synthesises carbohydrates.

(a) In plant cells, it has been demonstrated to produce pectic materials which are enclosed in vesicles, moves to the cell wall region. Here vesicle membrane fuses with the plasma membrane and the vesicle contents accumulate between the cell wall and the plasma membrane.

(b) Similar observations have been made in the formation of new cell plate at the time of cell division in plants. Abundant dictyosomes cut off vesicles, which move to the cell plate region and fuse together, releasing constituents of new cell plate.

4. Enzymes and acid have often been demonstrated to be active in dictyosomes. The most characteristic enzymes, associated with Golgi membrane, are those linking sugars with proteins to form glucoproteins, for example - mucopolysaccharide used by *Hydra* for attaching its base to the substratum.

5. Neutra and Leblond carried out autoradiographic studies to show that Golgi produces pectic substances. They proposed the following scheme for polysaccharide synthesis.

(a) Precursors enter the goblet cells from the capillaries of the vascular system.

(b) The amino acids are synthesised into proteins on the ribosomes of the ER.

(c) The proteins are then transferred to the cisternae of the Golgi complex.

(d) The simple sugar molecules go directly from the blood stream to the Golgi complex cisternae, where they are complexed with protein to form glycoprotein.

(e) The glycoprotein is then sulphated in the Golgi complex to form mucigen.

(f) The distal cisternae becomes rounded off and are transformed into mucigen granules.

(g) The mucigen granules are progressively displaced towards the apex of the cell. During this process, the mucigen undergoes further modification.

(h) The mucigen granules are secreted into the intestine by exocytosis during which the vesicle membrane fuses with the plasma membrane.

6. It takes an active part in the incorporation of substances from the environment.

7. Formation of Acrosome during Spermatogenesis: During the maturation of sperm, the Golgi plays a role in the formation of acrosome. In early stages Golgi appears as a spherical body, comprising cisternae arranged in parallel stacks and numerous small vesicles. These are always pinched off from the cisternae. As development proceeds, the Golgi becomes irregular in shape and large vacuoles are formed by dilation of cisternae sacs. In the centre of these large vacuoles, are present dense granule and the proacrosomal granules. These granules seem to be the enlarging vacuole. This vacuole along with granules approaches the anterior pole of the nuclear membrane constituting acrosomal granule. With the elongation of the spermatid, the acrosomal vesicle spreads over the nuclear surface and finally collapses with the nuclear membrane, forming a cap material. The acrosomal granule becomes the acrosome which lies at the apex of the nucleus and apparently comprises certain enzymes involved in the process of fertilisation.

8. Pigment Formation: In many mammalian tumour and cancer cells, the Golgi complex has been described as the site of origin of pigment granules (melanin). The Golgi complex has also been associated with pigment formation, in renal pigment epithelium of chick embryo, in the test cell in the ovary of the tunicate and in the oocytes of the Salamander.

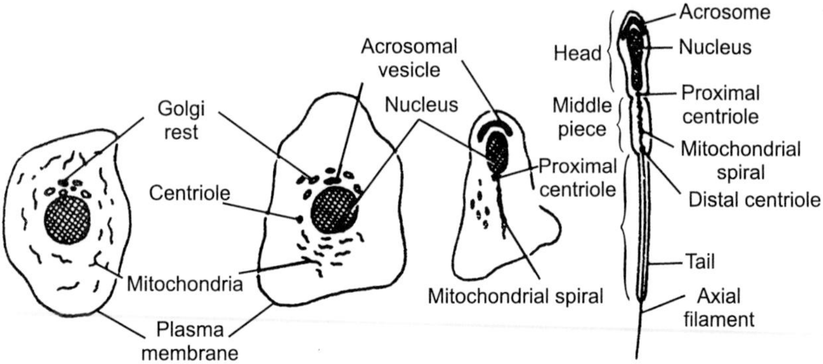

Fig. 5.3: Formation of acrosome from Golgi complex

5.3 LYSOSOMES

The lysosomes were first isolated in 1949 by **Christian de Duve** as a class of cell particles having centrifugal properties intermediate between those of mitochondria and ribosomes. These particles were rounded; dense bodies observed in hepatic cells and first called *pericanalicular dense* bodies. In 1955, **de Duve** renamed these bodies as lysosomes (lyso = digestive; some = body) because they contain digestive enzymes capable of lysis or digestion.

5.3.1 Occurrence and Origin of Lysosomes

The lysosomes may originate directly from the endoplasmic reticulum or from the Golgi membranes and then they associate with the vesicles formed by pinocytosis or phagocytosis. There is evidence that the lysosomes are also originated from the Golgi complex. The protein granules synthesised by the ribosomes are seen in the enlargements of endoplasmic reticulum. When these are budded off, they give rise to bodies having properties of lysosomes. They are also formed by pinocytic vesicles.

Lysosomes have been found both in animal and plant cells and in protozoa. In bacteria, there are no lysosomes, but the so-called *periplasmic space,* found between plasma membrane and the cell wall, may play a role similar to that of the lysosomes.

5.3.2 Structure of Lysosomes

It is difficult to identify the lysosomes under electron microscope because they have no characteristic shape or structure. They usually appear as a dense globular body and vary in size from 0.2 to 0.8 μ in diameter. They may be large in mammalian kidney (5μ) and exceedingly large in phagocytes. The lysosome is essentially a sphere of single lipoprotein membrane containing hydrolytic enzymes; which are involved in cellular digestion. If the enzymes are released, they can digest the cell. The membrane is impermeable to substrates of the enzymes contained in the lysosomes. The substances having molecular weight higher than 200 do not diffuse through the membrane. This indicates that lysosomal hydrolases have no direct access to cellular components which prevents uncontrolled digestion of the cell contents. Thus lysosomal membrane is resistant to the enzymes that it encloses, and entire process of digestion is carried out within the lysosome. The enzymes have destructive effects on the rest of the cell; hence the stability of the membrane is of great importance to the normal function of the cell. In fact, pathological conditions are known in which this membrane becomes more labile and permits exit of the enzymes causing destruction of the cells. Another interesting point is that most of the lysosomal enzymes act in an acidic medium.

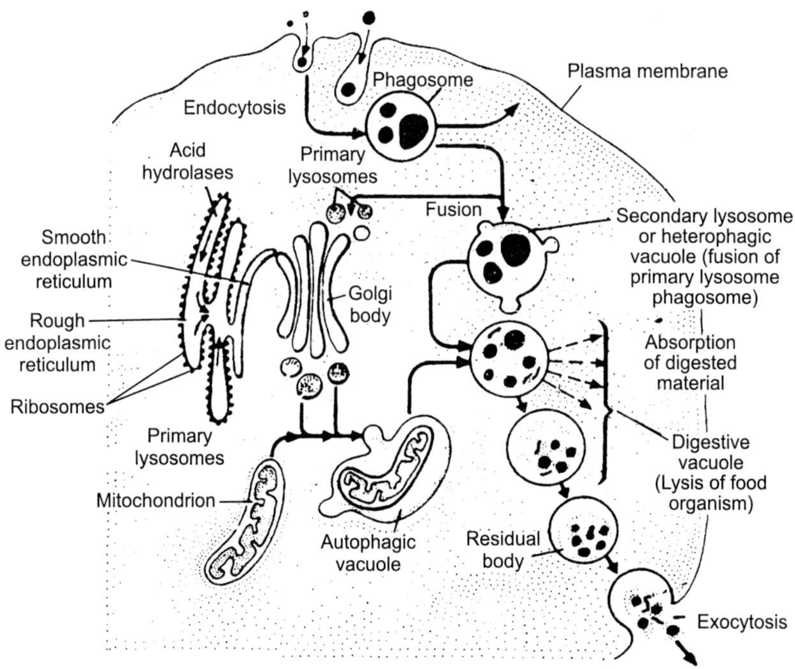

Fig. 5.4: Formation of lysosomes from Golgi Complex

The identification of the lysosomes is possible with the use of electron microscope and various cytochemical techniques. The most widely used procedure is the *Gomori* stain for acid phosphatase. Cytochemical reactions for β-glucuronidase, aryl sulfatase, N-acetyl, β-glucosaminidase and 5 bromo 4-chloroindoleacetate – esterase can also stain the lysosomes. Certain substances like neutral red, antimalarial drugs, vitamin A are taken up by lysosomes and may be used in their identification.

5.3.3 Types of Lysosomes

The lysosomes are polymorphic. They show irregularities in size and internal structure. Because of polymorphic nature, lysosomes are extremely dynamic and this can be hydrolysed after isolation of the lysosomes. When lysosomes are within the cell, they are surrounded by *multivesicular bodies,* smooth and coated vesicles and dense bodies. These bodies show fusion and fission events.

Four types of lysosomes are distinguished of which only the first is *the primary lysosome;* the other three may be grouped together as *secondary lysosomes.*

(1) Primary lysosomes: They are also called storage granules. Each primary lysosome is small body and its enzymatic content is hydrolyses by the ribosomes and accumulated in the endoplasmic reticulum. From these the enzymes penetrate into the Golgi region, in which the first acid phosphatase reaction takes place. The primary lysosome may be charged preferentially with one type of enzyme or the other.

(2) Digestive vacuoles: They are also called heterophagosomes. They are formed by phagocytosis or pinocytosis of foreign material by the cell. These are secondary lysosome which contains the engulfed material within a membrane and shows positive acid phosphatase reaction. This may be due to fusion with primary lysosomes. They contain enzymes and the material to be digested or already digested materials are present.

(3) Residual bodies: After digestion, the end products pass through the lysosomal membrane and are incorporated into the cell. The undigested material is left or if the digestion is incomplete the secondary lysosomes formed are called residual bodies. In some cells such as

Amoeba and other protozoa, these residual bodies are eliminated. But in other cells, they remain for long and may load the cell.

(4) Autophagic vacuoles: These secondary lysosomes are also called *cytolysosomes* or autophagosomes. These are special types of lysosomes in which they contain a part of the cell in the process of digestion (e.g. a mitochondria or portions of endoplasmic reticulum) Lysosomes regularly engulf bits of cytosol, which is degraded by a mechanism called *microautophagy.* During starvation, the liver cells show numerous autophagic vacuoles. This is a mechanism by which the cell can achieve the degradation of its own constituents which have undergone irreparable damage.

5.3.4 Lysosomal Enzymes

Lysosomes contain variety of enzymes. There are about 40 enzymes that have been isolated. Most of the enzymes function better under acidic condition (optimum pH – 5) and they are collectively called as *acid hydrolases.* Some of the enzymes found in lysosomes are shown in table 5.1.

Table 5.1: Some Acid Hydrolases of Lysosomes

	Enzyme	Natural Substrate	Source of Lysosomes
(a)	Nucelases Acid ribonuclease Acid deoxyribonuclease	RNA DNA	Many tissues of animals and plants.
(b)	Phosphatases Oligonucleotides Acid Phosphodiesterase	Most phosphate monoesters and phospho diesters	Many tissues of animals and plants
(c)	Proteases Cathepsin Collagenase Peptidases	Proteins Collagen Peptides	Animals Bone Animals and Plants

Contd...

(d)	Lipases Esterases Phospholipases	Fatty acids esters Phospholipids	Animals and plants
(e)	Glycosidases β-Galactosidase α-Glucosidase β-Glucoronidase	Galactosides Glycogen Glycogen Mucopolysaccharides Polysaccharides	Liver
	Lysozyme	Bacterial cell wall and kidney mucopolysaccharides	Egg white Saliva
	Hyaluronidase Arlysulphatase	Hyaluronic acids Chondroitin sulphates Organic sulphates	Liver Liver, Plants

The lysosomes degrade every biological macromolecule. A single lysosome need not contain all the enzymes. Many of the enzymes exist in multiple molecular forms.

5.3.5 Functions of Lysosomes

1. Intracellular digestion: The lysosomes contain enzymes and digest material intracellularly. Intracellular digestion is of two types: autophagy and heterophagy.

Autophagy is intracellular digestion in which enzymes of the cell digest the pieces of cytoplasmic material from the cell itself, i.e., digestion of endogenous material. Mitochondria are constantly removed by this process. Thus, there is renovation and turnover of cellular components.

Heterophagy means intake of exogenous material into the cell and its breakdown by lysosomal enzymes. Endocytosis is a phenomenon in which bulk of exogenous material is taken inside the cell. It includes two phenomena, pinocytosis, i.e. incorporation of liquid material and phagocytosis i.e. intake of solid material.

2. Extracellular digestion: The lysosomal enzymes are also released outside the cell and they hydrolyse the extracellular material. Lysosomal enzymes are also involved in bone erosion. Such process may occur in rheumatoid arthritis, a human disease in which the cartilage of the joints may be eroded by lysosomes.

3. Hormone secretion: The lysosomal proteases activate the secretion of thyroid hormones in the follicles.

4. Lysosomes are important in germ cells and fertilisation: The acrosome of spermatozoan is nothing but a special lysosome which contains protease and hyaluronidase. These enzymes are useful for penetration of sperm into egg and also for digestion of stored reserve materials.

5. Role in development processes: Many developmental processes involve the shedding or remodeling of tissues with the removal of whole cells and extracellular material. For example, the degeneration of the tadpole tail is produced by the action of cathepsins (i.e., proteolytic enzymes) contained in the lysosomes.

6. Lysosomes involved in human diseases and syndromes: In certain pathological conditions, such as rheumatoid arthritis, silicosis, and asbestosis (diseases produced by the inhalation of silica or asbestos particles) and gout (in which crystals of urate accumulate in the joints), there is release of lysosomal enzymes from the macrophages and acute inflammation of the tissues may lead to an increase in collagen synthesis (fibrosis). An acute release of lysosomal enzymes occurs in states of anoxia, acidosis and shock, and results in increased amounts of enzyme in the blood. The lipid-soluble vitamins (A, K, D & E) and the steroid hormones make the membrane more susceptible to rupture. On the other hand, cortisone, hydrocortisone and other drugs that have anti-inflammatory action, tend to stabilise the lysosomal membrane.

7. Storage diseases caused by mutations that affect lysosomal enzymes: Several congenital diseases have been found in which there is accumulation within the cells of substances such as glycogen or various glycolipids. These are called storage diseases and are produced by mutation that affects one of the lysosomic enzymes, involved in the catabolism of a certain substance. For example, in glycogenosis type II, the liver and muscles are filled with glycogen within membrane-bound organelles. In this disease α-glycosidase, the enzyme that degrades glycogen to glucose, is absent. There are about 20 diseases that occur due to lack of certain lysosomal enzymes.

8. Role in seed germination: Lysosomes are found in plant cells. In seedlings they are involved in the hydrolysis and in removal of protein and starch during germination.

POINTS TO REMEMBER

- The endoplasmic reticulum is found in almost all animal and plant cells except erythrocytes and prokaryotes.
- There are two basic morphological types of the endoplasmic reticulum (1) the rough endoplasmic reticulum or granular form and (2) the smooth endoplasmic reticulum or agranular form.
- The Golgi complex is supposed to be originated from the endoplasmic reticulum.
- The shape of Golgi is quite variable in somatic animal cells; even in the same cell, there is variation with functional stages.
- The small vesicles are 400 to 800 Å in diameter. They are intimately associated with the cisternae and may show continuity with them.
- During the maturation of sperm, the Golgi plays a role in the formation of acrosome.
- It is difficult to identify the lysosomes under electron microscope because they have no characteristic shape or structure.
- The lysosomes are polymorphic. They show irregularities in size and internal structure.
- Lysosomes contain variety of enzymes. There are about 40 enzymes that have been isolated.

EXERCISE

1. Give an account of ultrastructure of endoplasmic reticulum and describe its functions in brief.
2. Describe various types of ER. Enumerate the common and specialised functions of smooth endoplasmic reticulum.
3. Describe the position and structure of various types of ER. Explain its functions.

4. Differentiate between rough and smooth endoplasmic reticulum.
5. Mention the function of Endoplasmic Reticulum.
6. Give an account of the structure and functions of the Golgi body or Golgi apparatus. **(April 2014)**
7. Describe the structure and functions of lysosomes.
8. What are lysosomes? Describe origin, structure and functional significance.
9. Lysosomes are suicide bags in the cell. Explain
10. Define rough and smooth endoplasmic reticulum. Add a note on its physiological significance.
11. Describe the morphology, origin and chemical composition of Golgi complex.
12. Write short notes on:
    a. Morphology of Golgi complex
    b. Structure of Golgi complex
    c. Chemical composition of Golgi complex
    d. Functions of Golgi complex
13. Define Lysosomes. Describe their functions.
14. Define Lysosomes. How they can be regarded as polymorphic?
15. Differentiate
    a. Primary and secondary lysosomes
    b. Intracellular and Extracellular digestion

❖❖❖

6
CHAPTER

MITOCHONDRIA AND PEROXISOMES

CONTENTS

6.1 INTRODUCTION

Mitochondria are small granular, filamentous bodies generally known as "Powerhouse of the Cell". They are found in the cytoplasm of all aerobically respiratory cells of plants and animals, but not in bacteria where respiratory enzymes are located in the plasma membrane. Thus, they are associated with cellular respiration and are the sources of energy.

6.1.1 Morphology

Mitochondria vary in shape but are generally granular or filamentous. They may swell out at one end to become club-shaped. They may become vesicular by the appearance of central clear zone. Polymorphic forms may contain swellings at both ends.

The size of mitochondria is also variable. The average length of the mitochondria is 3 to 4 μ and the average diameter is 0.5 to 1.0 μ. The different physical and chemical factors such as pH, cell environment and the osmotic pressure influence the size and shape of mitochondria.

The number of mitochondria varies in different cell types. The number of mitochondria depends upon the metabolic activity of the cell. Cell with high metabolic activity have a high number of mitochondria, while those with low metabolic activity have a lower number. In giant amoeba *Chaos chaos* there may be 500,000 mitochondria. Large sea-urchin eggs have 13,000 to 14,000 while renal tubules have 300 to 400. They are fewer in green plant cells than in animal cells because some of their functions are taken over by chloroplast.

Generally, they are evenly distributed in the cytoplasm. In some cases, they are aggregated around the nucleus. In *Paramoecium* they are located beneath the cell surface. During cell division, they accumulate about the spindle and upon division of the cell, are distributed more or less equally among the daughter cell. In skeletal muscles they lie between the myofibrils.

6.1.2 Mitochondrial Formation or Biogenesis

There are four hypothesis / theories put forth to explain the origin of mitochondria:

(1) *De novo* origin of mitochondria.

(2) Mitochondrial formation from other intracellular structures.

(3) Formation by growth and division of pre-existing mitochondria.

(4) Prokaryotic origin of mitochondria or symbiont hypothesis.

(1) *De novo* origin of Mitochondria: *De novo* synthesis of mitochondria has been traced in a variety of cells. It is stipulated that mitochondria have developed from cytoplasmic vesicles by budding and formation of membrane. The buds thus formed grow in size, followed by compartmentalisation through vesicular formation, which ultimately develop into cristae. This hypothesis has little significance today because the observations supporting this theory were made with light microscope.

(2) Mitochondrial Formation from other Intracellular Structures: In this case, the mitochondria are originated by invaginations of the cell membrane. According to **Green** (1956) and **Robertson** (1964) the protuberance passes into the invagination of cell membrane to form double membrane lined tubule which enlarges into a vesicle. Soon from the inner membrane several outgrowths arise to form cristae, thus forming mitochondrion. Endoplasmic reticulum might have given rise to mitochondrion. A finger-like invagination of cytoplasmic matrix extends into the cavity of large endoplasmic reticulum cisterna.

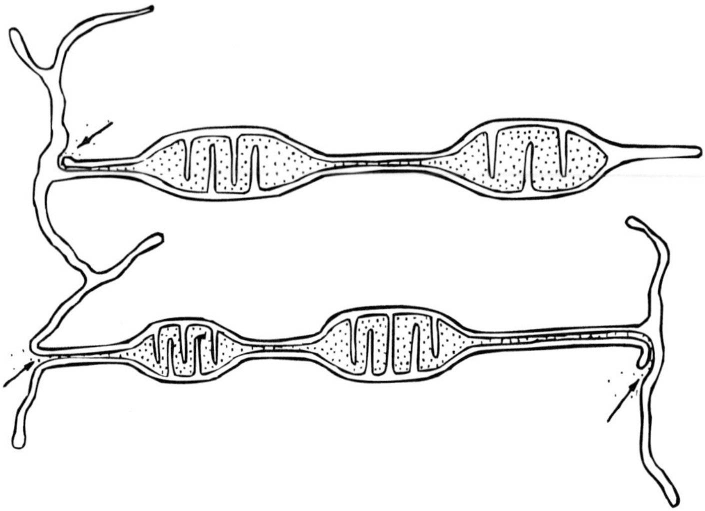

Fig. 6.1: A possible mechanism for the origin of mitochondrion

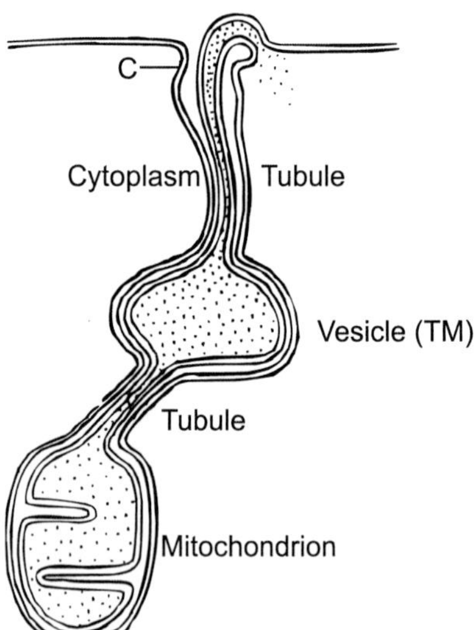

Fig. 6.2: A possible mechanism for the origin of mitochondrion from plasma membrane

(3) Formation of Growth and Division of Pre-existing Mitochondria: This was established by using labelled choline in choline-requiring mutant of *Neurospora crassa*. Labelled choline is incorporated in mitochondrial lecithin. The cells were examined after a three fold increase in population of mitochondria and each one of them contained labelled choline although they were not previously exposed to labelled choline. The presence of DNA, RNA, protein, phospholipid-synthesising machinery and metabolic pathways, all indicate their origin from pre-existing organelles. The mechanism first involves growth of the mitochondrial membranes and reproduction of organelles, followed by differentiation and compartmentalisation. During the phase of growth, the surface area of membrane increases, accompanied by elongation of the organelles and then partition by a few cristae forms. Ultimately, a constriction appears which separates the two daughter mitochondria.

(4) Prokaryotic Origin of Mitochondria or Symbiont Hypothesis: The old theory of the bacterial origin of mitochondria has been revived with recent discoveries of similarities in structure and physiology of bacteria and mitochondria.

According to this symbiotic hypothesis, primitive host cells which carried out anaerobic respiration by glycolysis were invaded by bacteria like parasites which respired aerobically through the Kreb's cycle and by oxidative phosphorylation. The invading organisms later developed an endosymbiotic relationship with the host cells and became the mitochondria. This theory is based on the several striking similarities between bacteria and mitochondria.

(i) Morphology: The general dimensions of bacteria and mitochondria are similar. Rod shaped bacteria are similar in shape to that of mitochondria.

(ii) Respiratory Chain: The inner membrane of mitochondria is similar to the bacterial plasma membrane with respect to the respiratory chain. The projections of plasma membrane (mesosomes) in some bacteria are similar to the cristae of mitochondria.

(iii) Chemical Composition: Lipid composition is similar in membranes of mitochondria and bacteria. Permease system is similar to that of bacteria.

(iv) Ribosomes: The small size of mitochondrial ribosomes (55-60S) is comparable to that of bacterial ribosomes (70S).

(v) Drug Sensitivity: Chloramphenicol drug inhibits protein synthesis in mitochondria and bacteria but not that of higher cells.

(vi) DNA Structure: In both cases DNA is circular. Circular DNA is found only in prokaryotes.

From these similarities one can easily conceive of mitochondria as being evolved from an ancient prokaryote.

6.1.3 Structure of Mitochondria

A typical mitochondrion is about 15,000 Å long and 5,000 Å in diameter. It is bound by two membranes, the outer membrane and inner membrane. The space between the two membranes is called intermembrane space. It is about 40-70 Å in width and filled with a watery fluid. The space enclosed by the inner membrane is called inner-membrane space or inner chamber. The inner space is filled with a matrix, which contains dense granules, ribosomes and mitochondrial DNA. These granules are considered to be the sites for binding divalent cations such as Mg^{++} and Ca^{++}.

The inner membrane is thrown up into a series of folds called mitochondrial cristae, which projects into the inner chamber. The cavity of the cristae is called the inter-cristae space, and is continuous with the intermembrane space.

The outer surface of the outer membrane and the inner surface of inner membrane are supposed to be covered with thousands of small particles. These sessile particles attached to the outer membrane are stalkless and are known as *subunits of Parson*. The particles on the inner membrane are stalked particles and they are called *subunits of Fernandez-Moran*. The particles on the inner membrane are considered to be involved in hydrogen transfer.

The outer membrane particles previously were considered to be hollow cylinders that are 60 Å long and 60 Å wide with a central hole of 20 Å in diameter.

The inner membrane particles consist of base piece, a stalk and a headpiece. The particles are spaced at a distance of 100 Å. The head piece is 75-100 Å in diameter and the stalk about 50 Å in length. It was formerly thought that each elementary particle contained all the enzymes for electron transport and oxidative phosphorylation. The particles were therefore called Electron Transfer Particles (ETP). Each particle is believed to consist of four complexes with complexes I and II situated in the base piece, complex III in the stalk and complex IV in the head piece.

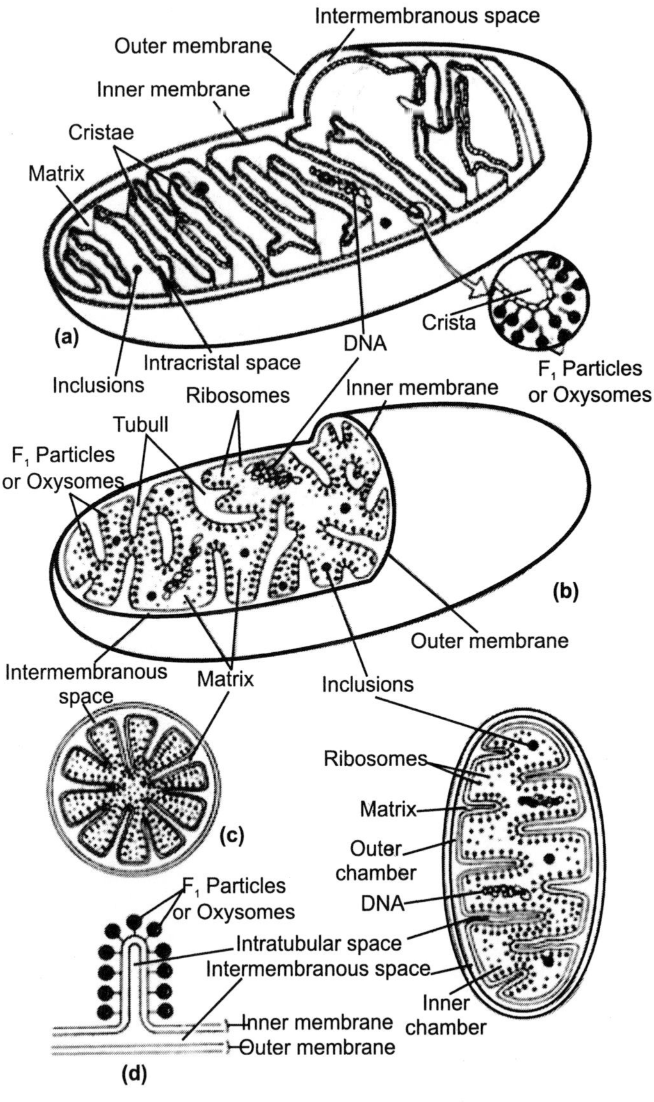

Fig. 6.3: Structure of Mitochondria

Chemical Composition

The mitochondrial membrane has a molecular organisation similar to that of plasma membrane. Each membrane is made up of two main substances besides the attached particles. These are proteins and lipids. The proteins are the main constituents and are about 4/5th of the dry weight of the membrane. In its usual form it is insoluble. The lipid forms 1/5th of the weight of the membrane and is found entirely in the form of phospholipids. Each phospholipid molecule is like a clothespin, the head bearing electrically charged atoms and the two legs are formed of long chain fatty acid. This molecule is insoluble. However, in aqueous solutions, they arrange themselves in a spherical form known as micelle (a group of phospholipid molecules), and becomes soluble. Cholesterol has been found to occur in some, not in all the membranes. Thus, the two membranes of mitochondrion differs in its chemical composition.

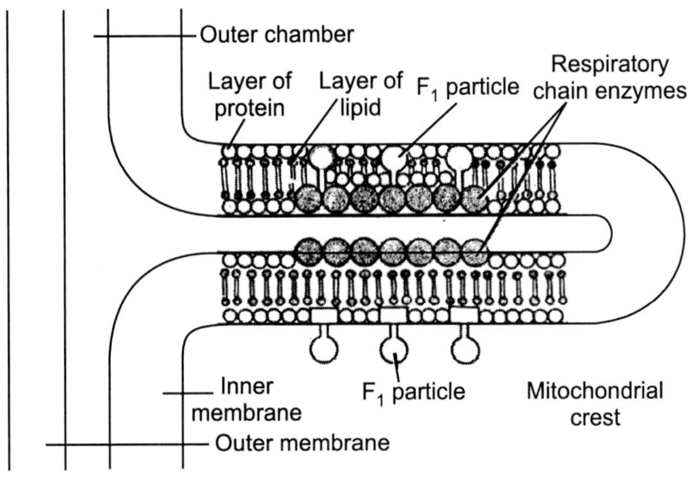

Fig. 6.4: Ultrastructure of mitochondrion to show the molecular organisation of Mitochondrial Crest

6.1.4 Functions of Mitochondria **(April 2014)**

The main mitochondrial function is to generate high energy ATP. It is because of this reason that they have been called **"the powerhouse of cell"**. For all the energy consuming processes of the aerobic cells, the energy in the form of ATP is made available by this cell organelle.

As the mitochondria can manufacture the ATP, the molecule having the potential energy is stored as high energy bond. They supply nearly all the required biological energy. Only the mitochondria are fully capable of converting pyruvic acid to carbon dioxide and water.

(1) Cell Respiration:

Mitochondria are the respiratory centres of the cell. Cell respiration can be divided into four steps:

(a) Glycolysis

(b) Oxidation of pyruvic acid

(c) Krebs /citric acid cycle

(d) Oxidative phosphorylation through hydrogen/electron transport system.

During the cell respiration, the organic substrate is oxidised to carbon dioxide and water in the presence of oxygen. One of the most important fuels is glucose. The overall equation for cell respiration in the presence of oxygen is:

$$C_6H_{12}O_6 + 6O_2 \longrightarrow 6CO_2 + 6\,H_2O + \text{Energy}$$

Under aerobic condition glucose metabolism takes place in four stages:

(a) Glycolysis: The breakdown of glucose of pyruvic acid is called glycolysis. Glycolysis can take place in the absence of oxygen (anaerobic condition) or in the presence of oxygen (aerobic condition). The enzymes for glycolysis are found in the soluble fraction of the cell, outside the mitochondria.

(b) Pyruvic Acid Oxidation: Under aerobic conditions, the pyruvic acid molecule is metabolised to a molecule of acetyl coenzyme A (acetyl CoA).

(c) Kreb's cycle: Each acetyl CoA molecule condenses with a molecule of oxaloacetic acid to produce a molecule of citric acid. After several steps, oxaloacetic acid is regenerated. Krebs cycle takes place under aerobic condition. The enzymes for the Kreb's cycle are located in the matrix of the mitochondrion.

(d) Oxidative Phosphorylation: It has been seen that pairs of hydrogen (2H) are liberated during aerobic glycolysis, oxidation of pyruvic acid and the Krebs cycle. These hydrogen pairs are passed down the hydrogen/electron transport chain (respiratory chain) and oxidative phosphorylation takes place. Three molecules of ATP are generated per pair of electrons passing down the electron transport chain. The enzymes for electron transport are located in the inner membrane of mitochondrion. The transfer of hydrogens/ electrons takes place through a complex chain of hydrogen donors-acceptors consisting of flavoproteins, coenzymes Q and cytochromes. The respiratory chain is arranged in five complexes (I to V). Oxygen is the final hydrogen acceptor.

(2) ATP Transport:

The ATP molecules produced as a result of cellular respiration accumulate in the mitochondria. The mitochondria collect at sites where energy requirement is high. As a result of membrane contraction and increase in internal hydrostatic pressure of the mitochondrion, water and ATP are squeezed out. This results in lowering of the ATP concentration, and the mitochondrial membrane relaxes. The thyroid hormone thyroxine causes the mitochondria to swell, while ATP brings about contraction.

6.2 PEROXISOMES AND GLYOXYSOMES

The cells of liver of vertebrates and protozoa, fungi, plants show the presence of certain membrane-bounded spherical bodies in close association with endoplasmic reticulum and mitochondria or chloroplast or both. These bodies measure 0.2 to 1.5 μm in diameter. These bodies possess a central granular or crystalloid core containing some enzymes. Such bodies are termed as **microbodies.**

Recent biochemical studies revealed that there are two types of microbodies, namely peroxisomes and glyoxysomes.

Both differ in their enzyme complement and in the type of tissue in which they are found.

6.2.1 Peroxisomes

Peroxisomes were discovered by **Tolbert** in 1969. These are small spherical bodies about 0.3 – 1.5 μm and bound by single unit membrane. Though similar in appearance to lysosomes, peroxisomes are not produced by budding from the Golgi apparatus. Instead, peroxisomes arise by the growth and division of existing peroxisomes. They resemble lysosomes in being filled with enzymes; however, peroxisome enzymes are oxidative of which catalase is the most characteristic. They are found in the leaves of higher plants in which photorespiration take place (in C_3 plants). In this process, green plants release CO_2 during day time. In animals, they are found in liver and kidney and protozoa. Peroxisomes have a matrix of fine granules in the central part of which there is a homogenous and opaque core. Glycolic acid oxidase, peroxidase, catalase and D-amino acid oxidase are found in high concentration in peroxisomes. The function of the core is that it is site of the enzyme urate oxidase in rat liver peroxisomes and the catalase in some plants.

Functions of Peroxisomes:

1. The catalase in peroxisomes converts the hydrogen peroxide (H_2O_2), a compound that is extremely toxic for living cells; to harmless water (H_2O) and oxygen (O_2).
2. Peroxisomes may also break down other toxic molecules that enter the cell, such as alcohol and many other drugs.
3. Peroxisomes within liver and kidney cells break down and detoxify about half the alcohol a person drinks, and other peroxisomes can produce a small amount of energy for the cell.
4. Peroxisomes participate in photorespiration with chloroplast and mitochondria.
5. They contain enzymes which participate in the oxidation of very long chain fatty acids, one chain of which contains more than 18 carbons.

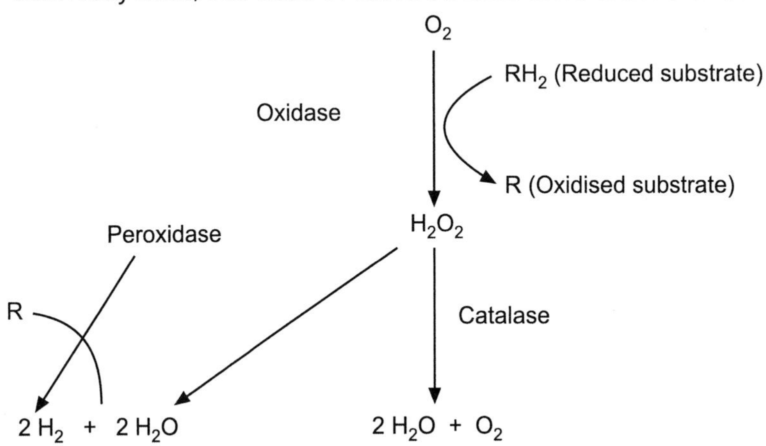

Fig. 6.5: Function of peroxisome

6. They participate in the synthesis of cholesterol and bile acids in liver and plasmogenes which are an important class of phospholipids in brain tissue in which one of the fatty acids in linked to the glycerol by an ether linkage rather than an ester linkage.
7. The enzyme luciferase, which generates the light emitted by fireflies, is also a peroxisomal enzyme.

6.2.2 Glyoxysomes

Glyoxysomes were described by **R.W. Briedenback** in 1967. Glyoxysome microbodies are found in the cells of yeast, *Neurospora* and oil rich seeds of many higher plants. Morphologically, they resemble peroxisomes, except that, their crystalloid core consists of dense rods (6.0 mµ in diameter). They contain the enzymes for fatty acid metabolism and gluconeogenesis i.e. conversion of stored lipid molecules of spherosomes of germinating seeds into the molecules of carbohydrates.

Functions of Glyoxysomes

1. **Fatty acid metabolism:** During germination of oily seeds, the stored lipid molecules of spherosomes are hydrolysed by the enzyme lipase (glycerol ester hydrolase) to glycerol and fatty acids. The phospholipid molecules are hydrolysed by the enzyme phospholipase. The long chain fatty acids which are released by the hydrolysis are then broken down by the successive removal of two carbon or C_2 fragments in the process of β-oxidation.

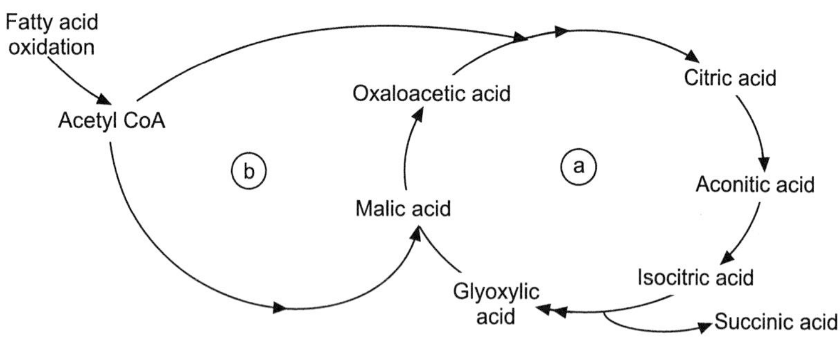

Fig. 5.6: Glyoxylate cycle (a) stand for enzyme isocitrate lyase and (b) stands for enzyme malate synthetase

2. **The enzymes** present in glyoxysomes are *isocitrate lyase* and *malate synthetase* and several essential enzymes of Kreb's cycle. During germination of seeds, the fat stored in the endosperm is transformed into carbohydrates by the way of glyoxylate cycle a modified form of Kreb's cycle.

The glyoxylate cycle does not occur in animal cells. They are unable to convert fatty acids into carbohydrates.

POINTS TO REMEMBER

- Mitochondria are small granular, filamentous bodies found in the cytoplasm of all aerobic animals and plants.
- The size of mitochondria is also variable.
- The number of mitochondria depends upon the metabolic activity of the cell.
- There are four theories put forth to explain the origin of mitochondria:
    a. *De novo* origin of mitochondria.
    b. Mitochondrial formation from other intracellular structures.
    c. Formation by growth and division of pre-exisiting mitochondria.
    d. Prokaryotic origin of mitochondria or symbiont hypothesis.
- Peroxisomes are small spherical bodies filled with enzymes that arise by the growth and division of existing peroxisomes.
- The enzymes of peroxisomes are oxidative in nature.
- Peroxisomes break down toxic molecules that enter the cell.
- Peroxisomes participate in photorespiration with chloroplast and mitochondria.

EXERCISE

1. Mitochondria are the "powerhouse of the cell". Discuss the statement.
2. Describe the structure and function of mitochondria.
3. Discuss the prokaryotic origin of mitochondria.
4. Write short notes on:
    a. Structure of Mitochondria
    b. Morphology of Mitochondria
    c. Chemical composition of Mitochondria
    d. Functions of Mitochondria **(April 2014)**

❖❖❖

7

CHAPTER

CELL DIVISION

CONTENTS

7.1 INTRODUCTION (April 2014)

The life cycle of a cell begins with the formation of daughter cells at the end of telophase. These daughter cells are smaller than the parent cell. Moreover, their DNA content is just half of the parent nucleus. These grow in size by synthesising new cytoplasmic and nuclear material till the total volume of each cell becomes four times of its original volume and the DNA content gets doubled by the replication of DNA. After this the cell is again ready to undergo division.

Thus, the cell cycle is an ordered set of events, culminating in cell growth and division into two daughter cells. It involves two phases.

1. The growth phase or non-dividing phase or interphase.
2. The mitotic phase (karyokinesis) or Division phase

Howard and Pele (1953) have divided cell cycle into four phases or stages: G_1, S, G_2 and M phase. The G_1 phase, S phase and G_2 phase are combined to form the classical **interphase**. Although the interphase appears to be a resting stage between cell divisions, it is actually a period of diverse activities. These interphase activities prepare the cell for the next mitosis phase. Cells however, spend most of their life-span in interphase.

G_1 Phase or Post mitotic Phase:

After the M-phase of previous cell cycle, the daughter cell enters the G_1 phase of interphase of the new cell cycle. G_1 - the resting phase is also called as the **first gap phase**, because during this phase no DNA synthesis take place. Recently, G_1 has been termed **first growth phase**, since it involves synthesis of RNA, proteins and membranes which leads to the growth of nucleus and cytoplasm of each daughter cell towards their mature size.

During this period the young daughter cell grows in size by synthesising cytoplasm. It synthesises and stores enzymes, mRNA, tRNA, ribosomes and proteins and also nitrogenous bases all of which are needed for DNA replication. During this phase chromosomes are extended to form slender fine threads and form an interwoven chromatin network.

The G_1 phase occupies about 30 - 50% time of the complete cell cycle. In mammalian cells whose life cycle is about 16 hours duration, G_1 phase is of about 5 hours. In non- dividing mammalian cells (for example, lymphocytes, nerve cells) the cell cycle stops at a point in this phase. This phase is known as the G_0 phase. This may be a temporary resting period or permanent. An example of the latter is a cell that has reached an end stage of development and will no longer divide (e.g. neuron). The nerve and heart muscle cells, become quiescent when they reach G0 phase but continue to perform their main functions for the rest of the organism's life.

Proteins synthesised during G_1 phase are:

i) Regulatory proteins which control various events of mitosis.

ii) Enzymes (i.e. DNA polymerase) necessary for DNA synthesis of the next stage.

iii) Tubulin and other mitotic apparatus proteins.

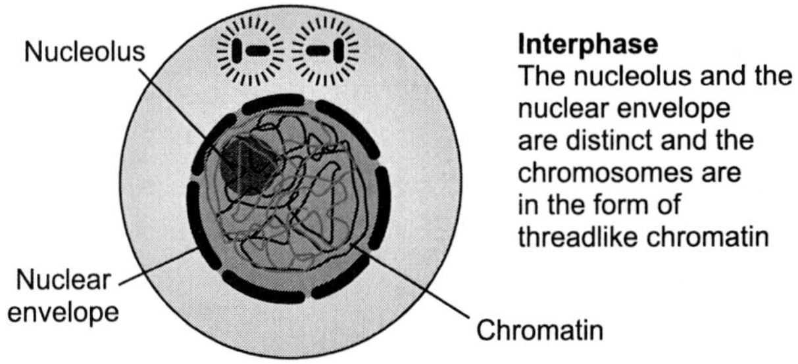

Fig. 7.1: Interphase

S-Phase (Phase of DNA Synthesis):

S-phase is the synthetic phase of interphase. During this phase replication of DNA and synthesis of histone proteins which are associated with DNA occurs. It occupies about 35 – 45 % time of the cell cycle.

This phase is an essential aspect of the cell cycle because replication ensures that each cell created by cell division has the same genetic make-up. During this phase, the cell growth continues along with the synthesis of a number of proteins and enzymes that are involved in DNA synthesis. Once DNA replication is complete the cell contains twice its normal number of chromosomes and becomes ready to enter the G_2 phase.

G_2 phase:

It is the second gap or growth phase of the interphase. During G_2 phase, synthesis of ribosomal RNA, messenger RNA and nucleolar RNA takes place. G_2 phase is the interval between the completion of DNA synthesis and the beginning of DNA segregation. It is the post DNA synthesis period, during which the cell makes preparations for the initiation of mitosis. This phase occupies only 10 – 20% duration of the cell cycle. In G_2 the cell continues to grow before it is split into two independent cells during mitosis.

Events of Interphase:

The interphase is characterised by the following events:

1. During G_1 the nuclear envelope remains intact. The chromosomes occur in the form of diffused, long, coiled single stranded structure.
2. In G_2 they have two strands or chromatids each. i.e. DNA amount becomes double due to accumulation of rRNA and ribosomal proteins in the nucleolus.
3. DNA duplication during S period will be at 4C value i.e. after DNA synthesis whereas it has only 2 C value prior to synthesis.
4. The synthesis of histone proteins occurs along with DNA, but some may be produced in G_1 period. The synthesis of RNA occurs during most of the cell cycle but it occurs to a lesser extent in S and M stages.
5. In animal cells, an interphase cell has two pairs of centrioles.

Division phase or Mitotic division:

This is the last phase of cell cycle. It is also called M-phase (Gr, mitos = thread) and occurs in somatic cells (eukaryotic cells). This phase gives rise to two identical daughter cells.

This phase is sequentially divided into four stages.

a) **Prophase stage:** In which condensation of chromosomes occur.

b) **Metaphase stage:** In this spindle formation takes place and chromosomes are arranged at the equator of the cell.

c) **Anaphase Stage:** In this chromosomes move towards the poles.

d) **Telophase stage:** This represents the formation of nucleus and cytokinesis takes place.

7.2 MITOSIS (April 2014)

All organisms that reproduce sexually develop from a single cell, the zygote, produced by the union of two cells, the germ cells or gametes. Development from zygote to the full grown stage takes place by the division of cells. In multicellular animals, two types of cells are present, namely *somatic cells and germ cells.* The somatic cells are responsible for the formation of different parts of the body and germ cells form the gametes - sperms and ova. Mitosis or somatic cell division is the multiplication of the somatic cells into daughter cells of equal size both containing the same number of chromosomes as the parent cell. The term mitosis (Greek; *mitos* - thread) refers to the threadlike appearance of chromosomes in the early cell division. At the time of mitosis nucleus becomes completely reorganised. The mitotic cycle is divided into series of consecutive phases known as prophase, metaphase, anaphase and telophase. The period between two mitotic cycles is called interphase.

Mitosis involves a series of complex changes in both the nucleus and cytoplasm. The division of nucleus is called karyokinesis and the division of cytoplasm is called *cytokinesis.*

(1) Prophase: It is the longest stage of mitosis and may last for several hours. During prophase, the nucleus enlarges and the chromosomes start to condense by a process of folding of the chromatin fibres. Prophase chromosome is composed of two coiled chromatids which are a result of the replication of DNA during S phase. Chromatids are united by a centromere and are called dyad. As prophase progresses, the chromatids become shorter and thicker and chromosomes are evenly distributed in the nuclear cavity. They migrate towards the nuclear envelope, leaving central clear empty space. At this point, each chromosome is composed of two cylindrical, parallel elements and they show primary and secondary constrictions. Other changes are the

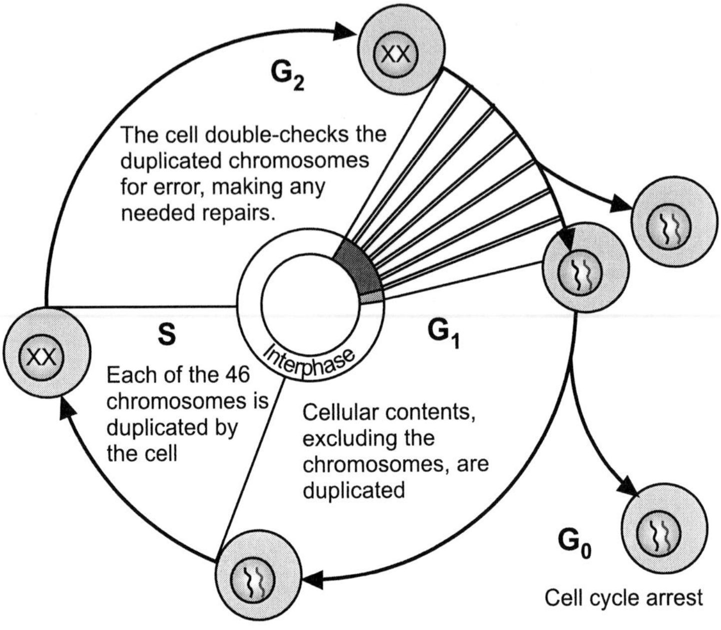

Fig. 7.2: Cell cycle and interphase

reduction in size of the nucleoli, and their disintegration within the nucleoplasm. At the end of prophase, the nuclear envelope is rapidly fragmented and disappears releasing nuclear material into the cytoplasm.

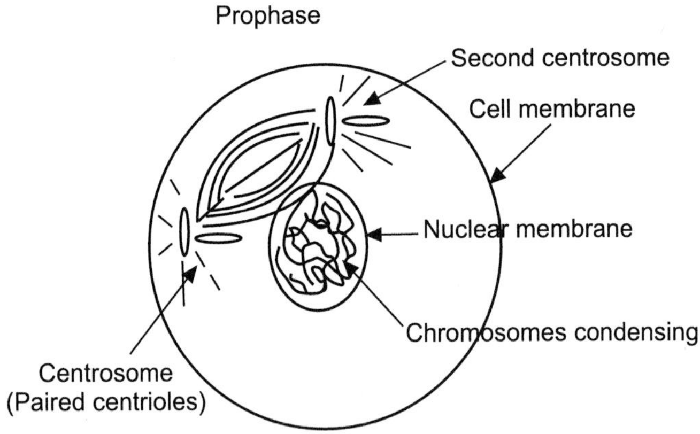

Fig. 7.3: Prophase

The most conspicuous change which occurs in the cytoplasm is the spindle formation. In the early prophase, there are two pairs of centrioles, each surrounded by the aster, composed of microtubules that radiate in all directions. The asters are cytoplasmic in origin. The two pairs of centrioles migrate towards opposite poles of the cell. They are pushed apart by the growth of the spindle fibres between them.

Pro-metaphase: Sometimes transition between prophase and metaphase is called pro-metaphase. This is a short period in which the nuclear envelope disintegrates and the chromosomes are in apparent disorder. There is no differentiation between cytoplasm and nuceloplasm. Such mitosis is called extranuclear mitosis or eumitosis.

(2) Metaphase: In this stage, the chromosomes reach the central or equatorial portion of the spindle. They become radially oriented at the equator to form the equatorial plate. Those fibres of the spindle that connect to the chromosomes are generally called the chromosomal fibres, and those that extend without interruption from one pole to the other are called continuous fibres. The centrioles lie on the equator of the spindle. Usually, the arms of the chromosomes lie on the equator of spindle. Occasionally, only the centromeres lie on the equatorial plane and arms are directed towards the poles.

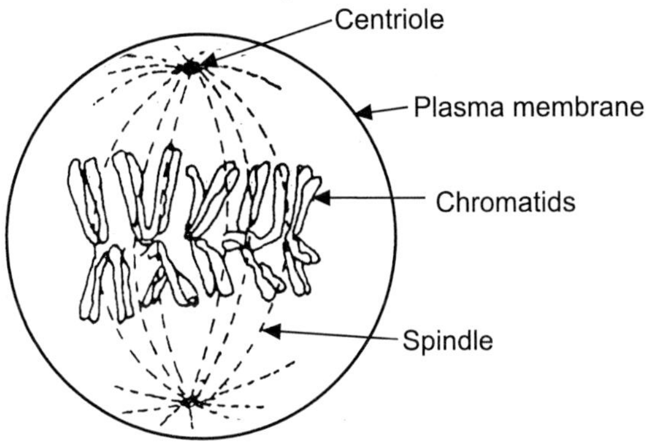

Fig. 7.4: Metaphase

Mitosis in which the spindle has centrioles and asters is called *astral or amphiastral* and is found in animal cells and some lower plants. Mitosis in which centrioles and asters are absent, is called *anastral* and is found in higher plants. Thus, centrioles and asters are not indispensable to the formation of the spindle, and in a certain way, the formation of spindle in astral mitosis is a mechanism that leads to the distribution of the centrioles between the two daughter cells. Spindles are of two types, *direct* and *indirect.* In the direct type, chromosomal fibres connect the chromosomes directly with the pole and in the indirect type; the chromosomes are connected with continuous fibres.

(3) Anaphase: The centromeres of the chromosomes arranged on equatorial plate, divide simultaneously as anaphase commences, and the two chromatids of each pair separate. They are now called daughter chromosomes. The centromeres move apart and the chromatids also start moving towards the poles. The forces behind chromosome movement are not yet known.

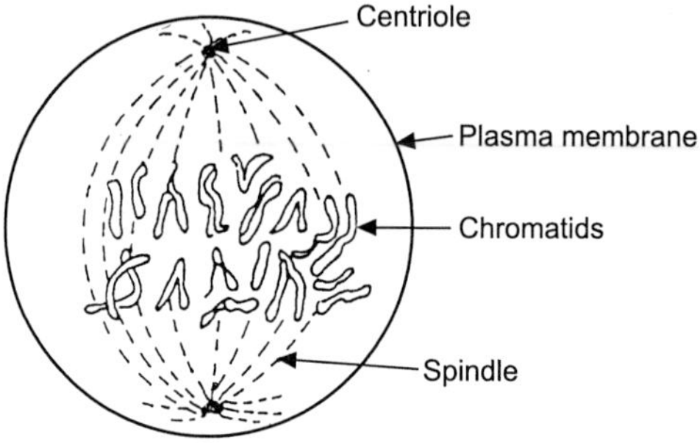

Fig. 7.5: Early Anaphase

It is suggested that the chromosomes are pulled by chromosomal fibres of the spindle. The chromosomes assume the shape of a V with equal arms if it is metacentric, or with unequal

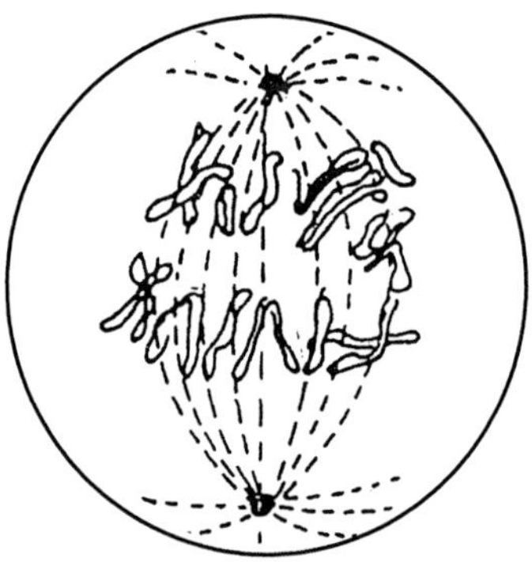

Fig. 7.6: Late anaphase

arms if it is submetacentric. During anaphase, the microtubules of chromosomal spindle shorten one-third to one-fifth of the original length. Simultaneously, the microtubules of the continuous fibres increase in length. Some of these stretched spindle fibres constitute the interzonal fibres.

(4) Telophase: The end of the polar migration of daughter chromosomes marks beginning of telophase (Gr. *Telo* = end). The chromosomes start to fold and become less and less condensed, thus telophase is the reverse of prophase. The spindle fibres disappear.

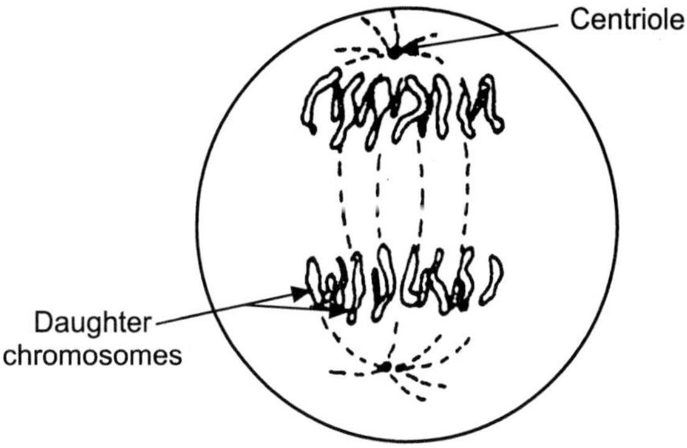

Fig. 7.7: Early telophase

The chromosomes gather into masses of chromatin that become surrounded by discontinuous segments of nuclear envelope made by the endoplasmic reticulum. Such segments fuse to make two complete nuclear envelopes of daughter nuclei.

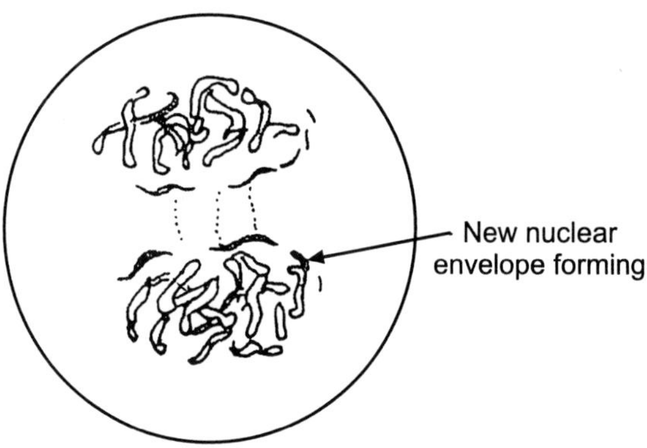

Fig. 7.8: Middle telophase

The nucleoli reappear at constrictions called the nucleolar organisers, in one pair of chromosomes. Each daughter cell gets the same complement of nuclei at the same site as the parent cell.

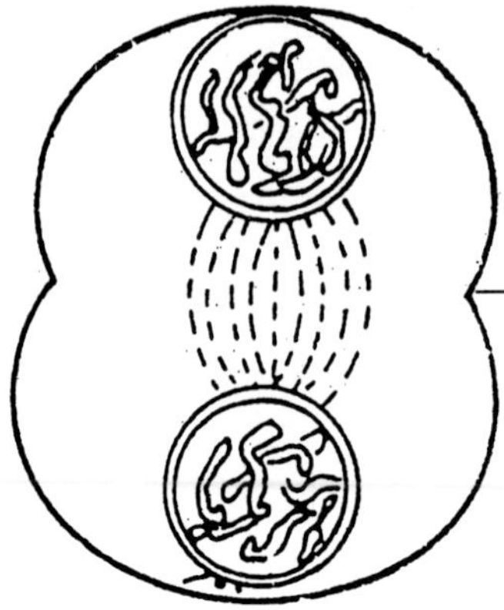

Fig. 7.9: Late telophase

Cytokinesis: It is the division of the cytoplasm of a cell following the division of the nucleus. It differs in plant and animal cells. Cytokinesis in animal cells proceeds by the process of furrowing at the equatorial region. The furrow is accentuated and deepened until the cell divides. The two cells may be equal in size or quite unequal. In plant cells, there is formation of cell plate between two groups of chromosomes. The cell plate grows from middle towards the periphery and finally joins the cell wall. This plate forms the membrane, separating the two cells. During cytokinesis, the cytoplasmic components are distributed including the mitochondria and Golgi complex.

Significance of Mitosis

1. **Equal distribution of chromosomes:** In the process of mitosis, the chromosomes are equally distributed to the two daughter cells. Thus constant number of chromosomes in all the cells of the body is maintained due to mitosis.

2. **Surface volume ratio:** Mitosis also maintains surface volume ratio of the cell. As the cell increases in size, the available surface area in relation to increased volume becomes less. Due to mitosis (or cell division) the cell becomes smaller in size and surface volume ratio is restored.

3. **Nucleoplasmic index:** A cell can't grow in size to a large extent without disturbing the ratio between the nucleus and cytoplasm. After a particular size has been reached, the cell divides to maintain the

nucleoplasmic index. Thus growth takes place by an increase in the number of cells, rather than by increase in the size of the cells.

4. Repair: Repair of the body takes place by addition of cells by mitosis. Cells are constantly sloughed off, dying and being replaced by new ones in the skin and digestive tract. When damaged tissues are repaired, the new cells must be exact copies of the cells being replaced so as to retain normal function of cells.

5. Regeneration: Some animals can regenerate parts of the body. Production of new cells in such cases is achieved by mitosis.

7.3 MEIOSIS (April 2014)

Meiosis is a special type of cell division found in organisms which reproduce by sexual reproduction. It occurs in the cells of gonads of animals and plants. The male and female gametes in sexually reproducing organisms fuse together and form the zygote. If the gametes had the same number of chromosomes as the somatic cells, then the zygote would have twice the diploid number of chromosomes. This number would go on doubling with each generation. But this never happens, as the chromosome number remains constant from generation to generation.

Meiosis is (Gr. meioum - to diminish) the mechanism that prevents the chromosome from doubling. By series of two meiotic divisions, the number of chromosomes is reduced to half which counteracts the effect of fertilisation. Thus fertilisation and meiosis are compensating events.

Meiosis essentially consists of two cell divisions which follow each other rapidly, while the chromosomes divide only once. The two divisions are known as the *first meiotic division* and *second meiotic division.*

Types of Meiosis:

There are three types of meiosis:

1. Zygotic or initial meiosis: In this type, fertilisation is immediately followed by meiosis, forming cell with haploid number of chromosomes. The zygote is the only **diploid** stage in the life cycle and it occurs in lower plants.

2. Gametic or terminal meiosis: It occurs in humans, all animal, and few lower plants. In the animal kingdom, this is the most common type of life history. Gametic meiosis is dominated by the diploid stage, with gametes formed directly by meiosis, and their fusion immediately regenerating the diploid condition.

3. Sporic or intermediate meiosis: It occurs in higher plants and not in animals. It takes place between fertilisation and the formation of gamete. In the diploid sporophyte meiosis takes place and forms megaspores and microspores. They undergo repeated mitosis and form gametophytes and certain cells differentiate to form gametes. Cells undergoing meiosis are often called *meiocytes.*

Following are the different stages of meiosis.

First Meiotic Division:

Prophase I
- Preleptotene
- Leptotene
- Zygotene
- Pachytene
- Diplotene
- Diakinesis

Pro-metaphase I

Metaphase I

Anaphase I

Telophase I

Second meiotic division

Interphase or interkinesis

Prophase II

Metaphase II

Anaphase II

Telophase II

7.3.1 First Meiotic Division

(1) Interphase:

The interphase preceding meiosis is important during this phase. DNA replication takes place, and it is confined to synthesis (S) phase of interphase.

(2) Prophase:

It is the first stage of meiosis and differs from prophase of mitosis. During this phase, nucleus starts increasing in volume due to hydration. It comprises mainly six substages, i.e. preleptotene, leptotene, zygotene, pachytene, diplotene and diakinesis.

(i) Preleptotene: It is the early prophase of meiosis. During preleptotene, the chromosomes are very thin and difficult to observe but only sex chromosomes may be seen.

(ii) Leptotene (Thin or slender thread) or Leptonema: During this period, the nucleus increases in size and the chromosomes become more distinct. Leptotene chromosomes look single rather than double. They appear as slender (thin) threads with bead-like thickenings called *chromomeres.* Because these beads are characteristic in size, number and position for a particular chromosome, they may be used as landmarks to identify a specific chromosome of an organism. Frequently, leptotene chromosomes have a definite polarization and form loops whose ends are attached to the nuclear envelope at points near the centrioles contained within an aster. This peculiar arrangement is often called the 'bouquet'. During leptotene phase, cytoplasm contains many polyribosomes, nucleolus increases and RNA and protein synthesis also increases.

(iii) Zygotene (mating thread) or Zygonema: During this stage, the chromosomes become shorter and thicker. The pairing or synapsis of homologous chromosomes begins. The homologous chromosomes undergo lengthwise pairing in which one chromosome has paternal origin (from male parent) called paternal chromosome, and other has maternal origin (from female parent) called maternal chromosome. This is the *bivalent stage.* Each pair of homologous chromosomes consists of four chromatids and this is sometimes referred to as *tetrad.* The pairing may begin at the centromere and proceed towards the end called *procentric pairing.* If it begins at the ends and proceeds towards the centromere, it is called *proterminal pairing.* Pairing is highly specific and involves the formation of a special structure that can be observed under the electron microscope and called *synaptonemal complex.* It lies between two homologous chromosomes as they do not fuse completely. This complex is composed of two lateral components or arms and a central or medial element. Each lateral component is shared by the two sister chromatids of homologue.

Morphology of the synaptonemal complex is very similar in plant and animal cells. In cross-section, it appears flattened ribbon-like structure. The lateral arms are formed of electron dense coarse granules or fibres. These arms are joined to the adjacent chromosomes by fine fibrils.

The synaptonemal complex is the structural basis for pairing and synapsis of meiotic chromosomes. The function of this complex is to stabilise the pairing between the homologues and to facilitate recombination.

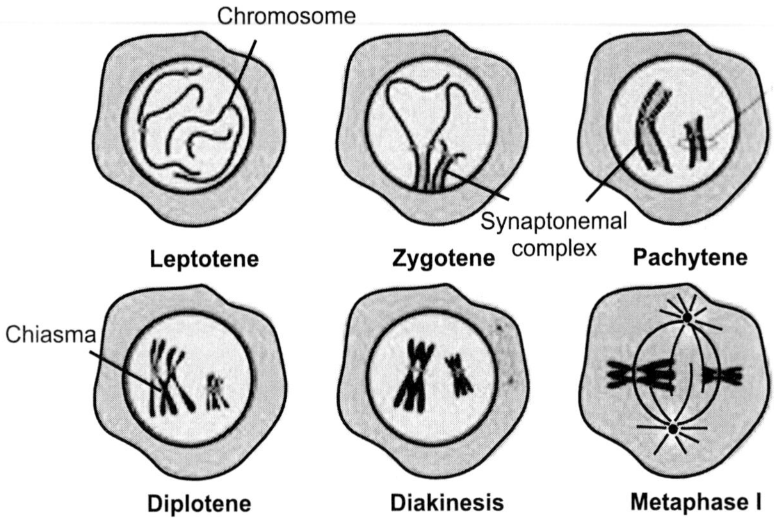

Fig. 7.10: Meiotic division stages of prophase I

(iv) Pachytene (Thick thread) or Pachynema: During this, the pairing of chromosomes is completed and it is a longer substage of prophase I. The chromosomes contract longitudinally so that threads become shorter and thicker. The chromosomes are associated in *bivalent* or *tetrads* composed of two homologous chromosomes in close longitudinal union and four chromatids. Each chromatid has its own centromere or kinetochore. Thus in a tetrad, first there are four centromeres, two homologous and two sister centromeres. During the first meiotic division, the centromeres of the two homologous chromatids behave as a functional unit. The chromatid is the unit of crossing over. During pachytene, two chromatids belonging to different homologous chromosomes undergo one or more transverse breaks at the same level. The breaks never occur between two sister chromatids. The break is followed by interchange and fusion of broken ends between two homologous chromosomes. This is called *crossing over*. Thus each bivalent consists of two chromatids which are unbroken and two chromatids whose broken ends have been interchanged.

(v) Diplotene or diplonema: During this stage, the intimately paired chromosomes repel each other and begin to separate. But this separation is not complete, because the homologous chromosomes are held together at one or more points where breaks and fusion have occurred. These points are called *chiasmata*. Chiasmata are generally regarded as the sites where the phenomenon of crossing over or recombination takes place. With few exceptions, chiasmata are found in all plants and animals.

At least one chiasma is formed for each bivalent. Their number is variable, some chromosomes show one chiasma while others show several. During diplotene, the four chromatids of the thread become visible and the synaptonemal complex disappears. A chiasma formed at the ends of chromosomes is called *a terminal chiasma* and if formed along the lengths of chromosomes it is called *interstitial chiasma.*

Fig. 7.11: Chiasmata formation

At the end of diplotene, the chiasma begins to move along the length of the chromosome from the centromere. This displacement is called *terminalisation*. The terminal chiasma slips off the ends of the chromosomes and its position is taken up by an interstitial chiasma. The degree of terminalisation is expressed by the terminalisation coefficient (T).

$$T = \frac{\text{Number of terminal chiasma}}{\text{Total number if bivalents (or nuclei)}}$$

The frequency of chiasma (F_9) is the average number of chiasma in a bivalent or in bivalents of a nucleus.

$$F_9 = \frac{\text{Total Number of terminal chiasma}}{\text{Total number of bivalents (or nuclei)}}$$

The rotation is well marked in bivalent having one chiasma. The arms of the bivalent rotate through 180° and form a cross. If there are two chiasma, a bivalent loop is formed. With more than two chiasma, a series of loops are formed. Rotation is brought about by same forces of repulsion which bring about terminalisation.

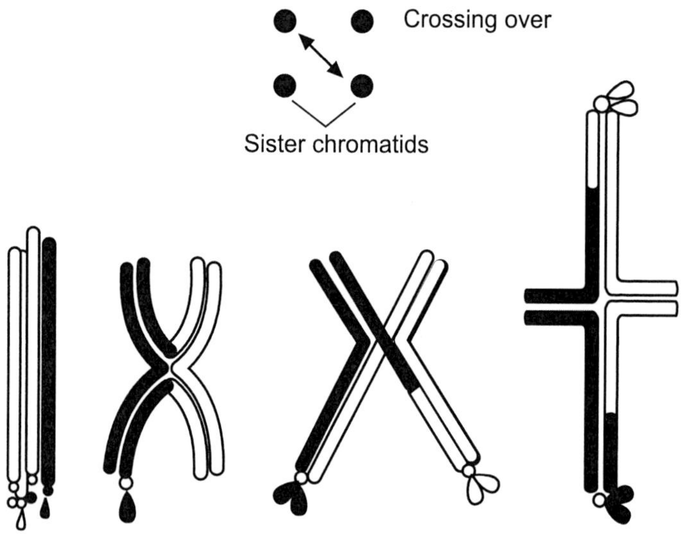

Fig. 7.12: Crossing-over, chiasma formation, terminalisation and rotation

(vi) Diakinesis: In this phase, the chromosomes again contract. The tetrads (bivalents) are more evenly distributed in the nucleus and nucleolus disappears. During this period, number of chiasmata diminishes. By the end of diakinesis, the homologues are held together only at their ends.

(2) Pro-metaphase I:

In this phase, condensation of the chromosomes reaches its maximum. The nuclear membrane disintegrates and disappears. Spindle formation begins.

(3) Metaphase I:

In this phase, the chromosomes become attached at the equator. The spindle is formed and the spindle fibres become attached to the centromeres of two homologous chromosomes. The two centromeres of each bivalent lie on opposite sides of equatorial plate. Metaphase I of meiosis differs from those of mitosis as there are bivalents, and each bivalent consists of two centromeres and they are being pulled towards the poles.

(4) Anaphase I:

In anaphase I, the sister chromatids of each homologue, united by centromere move towards the poles of the cells. The short chromosomes separate rapidly, while the separation of long chromosomes is delayed due to interstitial chiasmata. By recombination, segments are transposed between two of the chromatids of each homologue. In each homologue chromosome, one chromatid is unchanged, while the other has undergone mixing of maternal and paternal sections. There is no division of centromeres as in anaphase of mitosis.

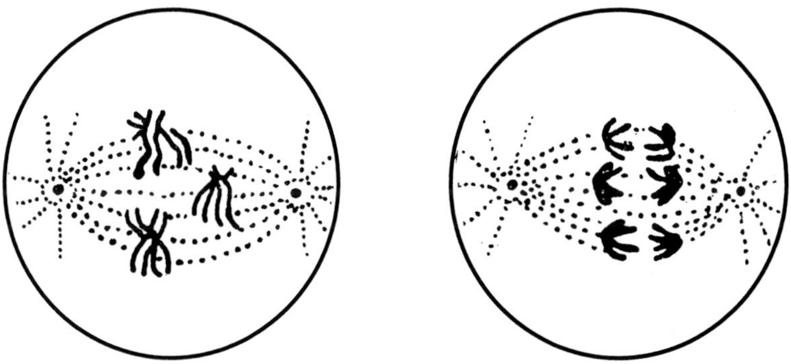

Metaphase I Anaphase I

Fig. 7.13: Metaphase I and Anaphase I

(5) Telophase I:

It begins when the chromosomes reach their respective poles. Chromosomes may remain in a condensed state for sometime. Later they undergo, depolaralisation and become elongated. Nuclear membrane is reformed and spindle fibres do not disappear completely. By cytokinesis in animal cells, two daughter cells are formed. In most plant cells, daughter cells are formed by forming cells plate between two groups of chromosomes. In other plant cells, cytokinesis takes places after completion of both the meiotic divisions.

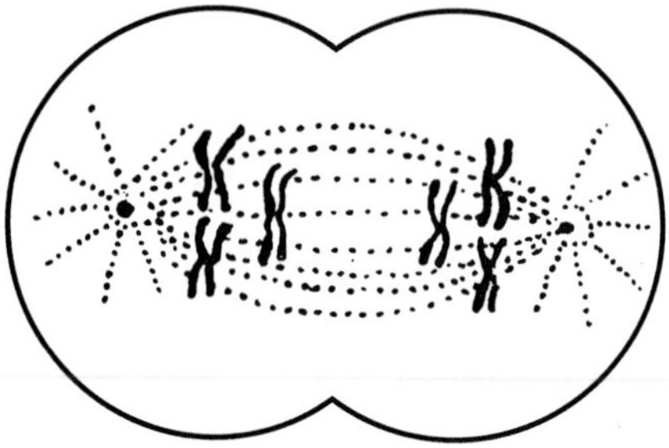

Fig. 7.14: Telophase I

7.3.2 Second Meiotic Division

(1) Interphase II or Interkinesis:

The second meiotic division is similar to mitosis; however, there is no DNA duplication. After telophase I, two haploid daughter nuclei or cells sometimes undergo typical resting stage as in mitosis. This intervening stage between telophase I and beginning of prophase II is called *interphase II* or *interkinesis*. This differs from interphase I and interphase of mitosis as there is no DNA duplication.

Fig. 7.15: Interkinesis or Interphase II

(2) Prophase II:

This stage is of shorter duration and much simpler than prophase I. Chromosomes undergo shortening and thickening. The sister chromatids have already been separated. The two chromatids of each chromosome remain separate except at the centromere. Spindle formation takes place in prophase II along with disappearance of nuclear membrane.

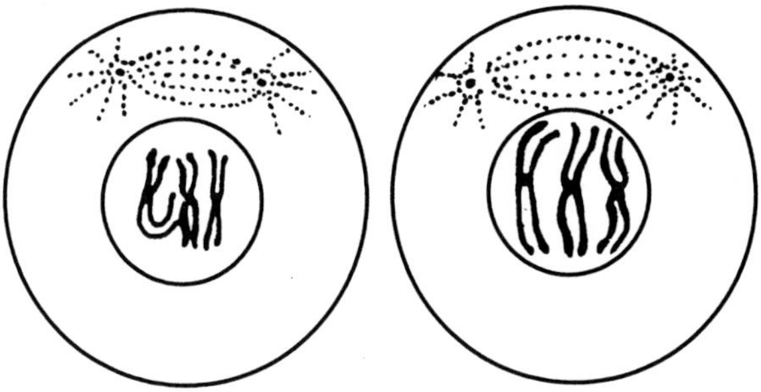

Fig. 7.16: Prophase II

(3) Metaphase II:

At this stage, chromosomes become arranged on the equatorial plane. The centromere of each chromosome splits longitudinally forming two centromeres. At the end of metaphase II, centromeres are connected to the spindle fibres from respective poles.

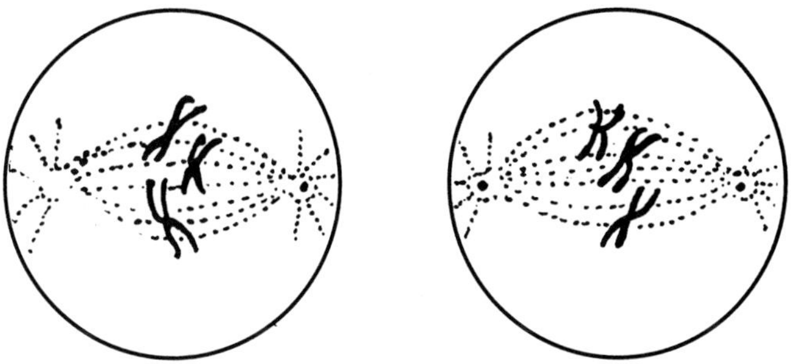

Fig. 7.17: Metaphase II

(4) Anaphase II:

It begins with the movement of chromatids to the opposite poles. They are pulled by contraction of spindle fibres.

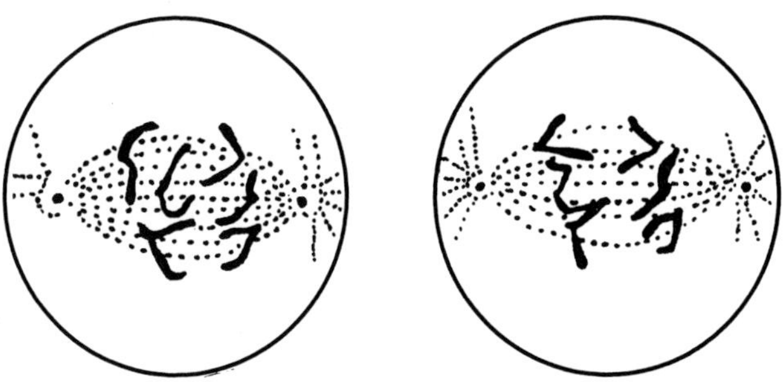

Fig. 7.18: Anaphase II

(5) Telophase

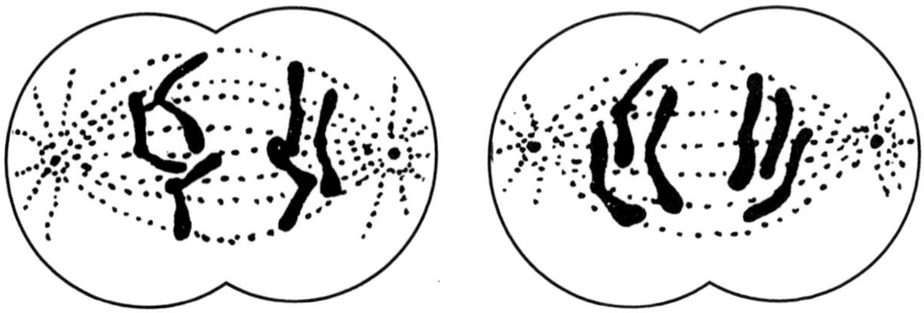

Fig. 7.19: Telophase II

The movement of chromatids to opposite poles ultimately results in the formation of nucleus at each pole. The nucleus, centriole and the chromosomes return to the interphasic condition. From two haploid nuclei at the end of meiosis I, four haploid nuclei are formed after the completion of meiosis II. Thus, meiosis II is an equational division.

Cytokinesis

Cell walls are formed between the daughter nuclei after the completion of telophase I. In some species, cell walls develop simultaneously at the end of the second division, while in other species, a cell wall is formed at the end of meiosis I and then successively at the end of meiosis II. In both the cases, four cells with haploid number of chromosomes in their nuclei are formed.

Significance of Meiosis

Meiosis is a very important event in the life cycle of sexually reproducing organisms. It is opposite of fertilisation. In fertilisation, the chromosome number is restored back to the original one, whereas meiosis reduces the chromosome number to half. The chromosomes in a species remain constant because of meiosis. For example, in human being, the somatic cells contain 46 chromosomes. If there will be no meiosis during gamete formation, the resultant gametes will carry 46 chromosomes. When the male and female gametes fuse, the zygote will possess 92 chromosomes. In the next generation, the chromosome number will be doubled to 184. But such situation does not happen due to meiosis; and hence each gamete possesses haploid number chromosomes.

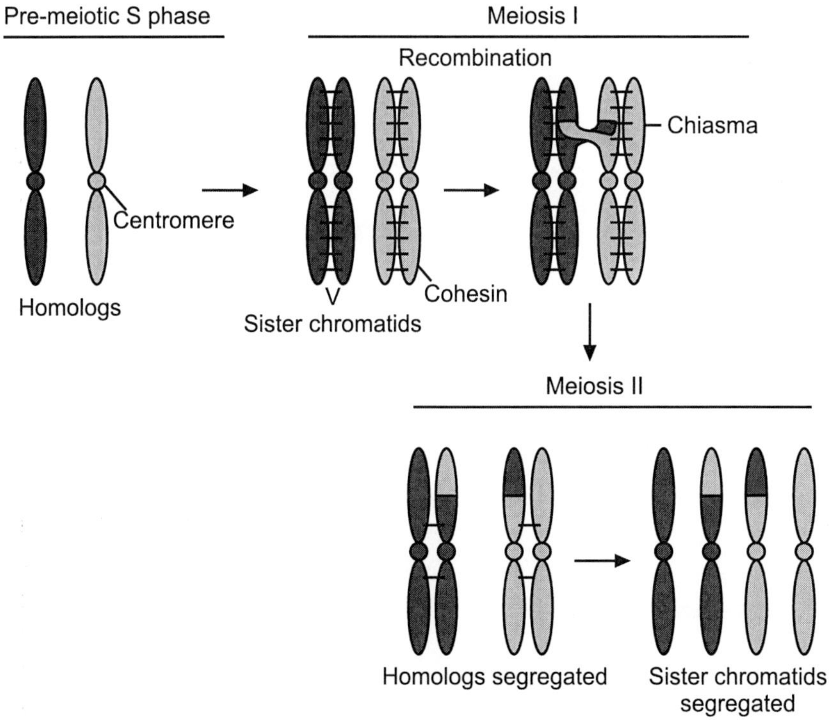

Fig. 7.20: Daughter cells

Secondly, there is reshuffling of genetic material during crossing over. This gives an opportunity for the exchange of genes between homologous chromosomes.

The separation of chromosomes at anaphase is at random, so the two nuclei formed at the end of meiosis I, are neither purely maternal nor paternal but random combination of both.

Random separation of chromosomes and crossing-over are responsible for the variations found in sexually reproducing organisms. Variations provide raw material for evolution.

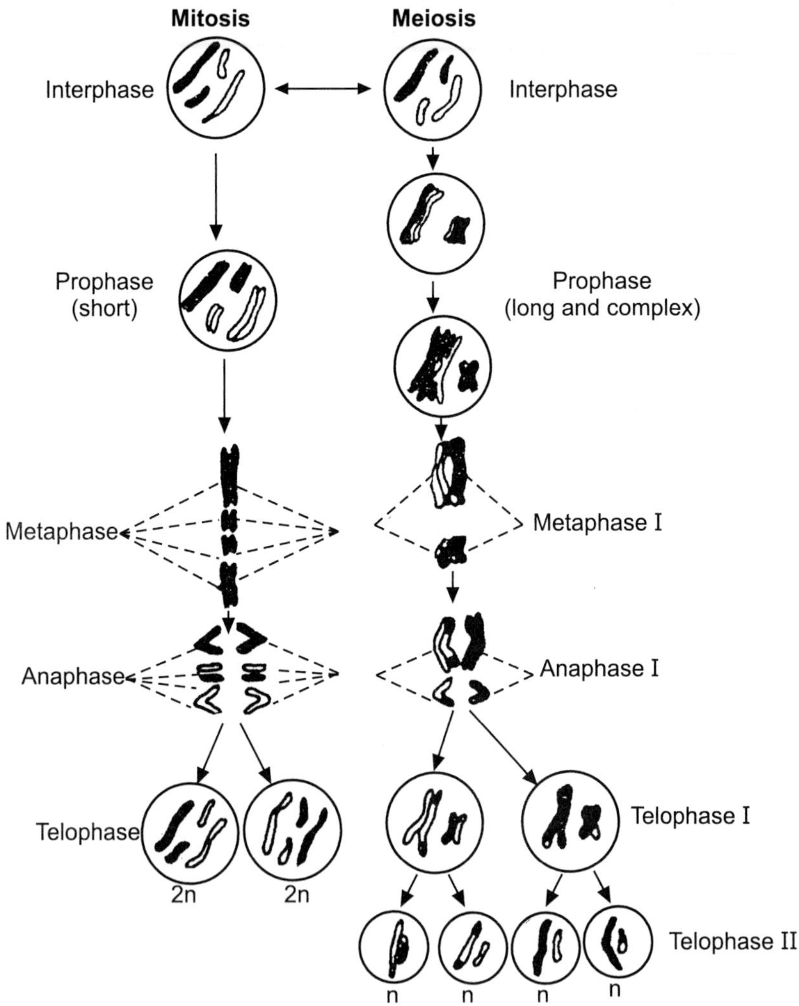

Fig. 7.21: Comparative account of mitosis and meiosis

Table. 7.1: Comparison of Mitosis and Meiosis

	Mitosis	Meiosis
1.	This division occurs only once	It involves two cell divisions, the first and the second meiotic division
2.	It occurs only in somatic cells	Meiosis occurs only in the germ cells
3.	It occurs in both sexually and asexually reproducing organisms	It occurs only in sexually reproducing organisms
4.	In mitosis DNA synthesis occurs in the S phase.	In meiosis there is premeiotic DNA synthesis which is much longer than that in mitosis
5.	In mitosis every chromosome behaves independently	In meiosis the homologous chromosomes become mechanically repeated during the first meiotic division.
6.	Mitosis is short process of 1 to 2 hrs.	Meiosis is long process
7.	In mitosis, the genetic material remains constant	In meiosis, genetic variability is one of the main consequences
8.	In mitosis DNA replication occurs during interphase – I	DNA replication occurs during interphase I but not interphase II
	Prophase	
7.	The prophase is of short duration usually of few hours	The duration of prophase is long and takes several days
10.	Prophase is very simple	Prophase is complicated and divided into many substages

Contd...

11.	Chromosomes are double structures and consist of two chromatids united by centromere	The two homologous chromosomes form bivalents or tetrads having four chromatids and two centromeres
12.	There is no exchange of chromosomes during prophase	There is exchange of segments of chromatids of homologous chromosomes
13.	The arms of the chromatids are close to one another	The arms of chromatids are separated widely in prophase II
	Anaphase	
14.	The centromeres lie towards the equator and arms directed towards the poles.	In metaphase I, the centromeres are lined up in two planes which are parallel to each other and their arms are directed towards equator.
15.	Metaphasic plate consists of chromosome pairs	Metaphasic plate is made of paired chromosome pairs.
16.	The centromeres divide during anaphase.	No centromere division occurs during anaphase II.
17.	Chromosomes separate	Short chromosomes separate rapidly but long chromosomes show delayed separation.
	Telophase	
18.	In telophase, spindle fibres disappear completely.	Spindle fibres do not disappear in telophase I
17.	Nucleoli reappear at telophase.	Nucleoli do not reappear during telophase I
20.	Cytokinesis always takes place after each division	Cytokinesis does not necessarily occur after telophase I, but it normally extends to the IInd meiotic division.

POINTS TO REMEMBER

- The cell cycle is an ordered set of events, culminating in cell growth and division into two daughter cells.
- The cell cycle involves two major phases-growth phase or non-dividing phase or interphase and mitotic phase (karyokinesis) or Division phase.
- The G_1 phase occupies about 30 – 50% time of the complete cell cycle.
- The S-phase is the synthetic phase of interphase which involves replication of DNA and synthesis of histone proteins which are associated with DNA.
- It occupies about 35 – 45% time of the cell cycle.
- G_2 phase is the second gap or growth phase of the interphase which involves synthesis of ribosomal RNA, messenger RNA and nucleolar RNA.
- Mitosis is the multiplication of the somatic cells into daughter cells of equal size both containing the same number of chromosomes as the parent cell.
- During prophase, the nucleus enlarges and the chromosomes start to condense.
- The chromatids become shorter and thicker; chromosomes are evenly distributed in the nuclear cavity.
- At the end of prophase, the nuclear envelope is rapidly fragmented and disappears releasing nuclear material into the cytoplasm.
- During metaphase, the chromosomes are arranged radially at the equator to form the equatorial plate.
- During anaphase, the chromosomes arranged on equatorial plate divide and the two chromatids of each pair separate.
- During telophase, the chromosomes become less and less condensed and the spindle fibres disappear.
- Cytokinesis is the division of the cytoplasm of a cell following the division of the nucleus.

EXERCISE

1. Write an essay on cell cycle.
2. Describe somatic cell division or mitosis in an animal cell with suitable diagrams.
3. Give neat and labelled diagrams to explain mitosis.
4. Give a detailed account of the meiotic cell division and explain its significance.
5. Describe process of meiosis and compare it with mitotic cell division.
6. Describe significance of prophase-I of meiosis.
7. Differentiate between diplotene and pachytene.
8. Describe significance of mitosis and meiosis.
9. Explain the term cell cycle. Explain the different phases of the cell cycle.
10. Differentiate between mitosis and meiosis.
11. Write short notes:
    a. Significance of meiosis
    b. Cell cycle
    c. Cytokinesis
    d. Zygotene
    e. Pachytene
    f. Diplotene
    g. Significance of mitosis
    h. Anaphase
    i. Interphase
    j. Metaphase
    k. G_0 phase
12. Define the terms: cell cycle and mitosis. Name the stages of cell cycle. **(April 2014)**

13. What is Mitosis? Describe the different stages of mitosis.
14. Define mitosis and add a note on cell cycle.
15. What is meiosis? Add note on its significance.
16. Mention the similarities and differences between mitosis and meiosis.
17. How is meiosis necessary to sexual reproduction? What are the basic differences between mitosis and meiosis?
18. Describe the first meiotic division in brief.
19. Describe the second meiotic division and give its significance.
20. What is meiosis? Describe in detail the process of meiosis.

(April 2014)

❖❖❖

MODEL QUESTION PAPER

Total Marks : 35 **Duration : 2 Hours**

Note :

(a) Question 1 is compulsory.
(b) Solve any three questions from Q.2 to Q.5.
(c) Questions 2 to 5 carry equal marks.

1. Attempt any five of the following : (5)

(a) Define the cell. Describe the various branches of cell biology.
(b) Explain prokaryotic cell and eukaryotic cell.
(c) Differentiate active and passive transport.
(d) Define pinocytosis and phagocytosis.
(e) Lysosomes are suicide bags in cell. Explain.
(f) Mitochondria are the 'power house of the cell'. Explain.

2. (A) Describe the ultra structure of a prokaryotic cell. or
Describe the ultra structure of eukaryotic cell. (6)
(B) Give a brief account of fluid mosaic model of plasma membrane. (4)

3. (A) Give an account of ultra structure of endoplasmic reticulum and describe its functions in brief. or
Give an account of ribosomes and their functions. (6)
(B) Describe the structure and functions of lysosomes. (4)

4. (A) Give an account of the ultra structure and functions of mitochondria. or
Describe morphlogy, structure and functions of golgi complex. (6)
(B) What is Mitosis? Describe different stages of mitosis. (4)

5. Write short notes on any four of the following : (10)

(A) Functions of nucleus
(B) Euchromatin and heterochromatin
(C) Significance of meiosis
(D) Micrometry
(E) Desmosomes
(F) Applications of electron microscope.

❖❖❖